RESILIENT **COMMUNITIES** ACROSS **GEOGRAPHIES**

RESILIENT
COMMUNITIES

ACROSS

GEOGRAPHIES

EDITED BY

Sheila Lakshmi Steinberg
Steven J. Steinberg

Esri Press
REDLANDS | CALIFORNIA

Esri Press, 380 New York Street, Redlands, California 92373-8100
Copyright © 2021 Esri
All rights reserved.
Printed in the United States of America
25 24 23 22 21   2 3 4 5 6 7 8 9 10

ISBN: 9781589484818

Library of Congress Control Number: 2020947460

For purchasing and distribution options (both domestic and international), please visit esripress.esri.com.

Contents

Foreword

Resilient communities—a worthy goal that may never be fully achieved. To be truly resilient, a community must be prepared to minimize the impact of various challenges. It can respond to threats to its ideal way of life quickly, and its infrastructure and residents will rebound promptly in the aftermath of crisis. Since there will always be new stresses, pressures, and disasters with which to contend, resiliency is perhaps less a goal than a process we must consistently strive to improve.

The Steinbergs have put together an insightful text illuminating this process through theory and application of spatial methods to cope with and improve our reactions to environmental and social changes. The book is crosscutting in its scope and delves into an interesting breadth of topics, from ecosystem resilience to climate adaptation to urban and cultural resilience. Through this range of examples, readers will be able to discern patterns of theoretical frameworks, analytical methods, and GIS applications that will undoubtedly be relevant to their own fields of interest.

Focusing on spatial resilience makes sense for this book and for this time in our history. This Digital Age in which we now live is characterized by data availability, information transfer, and democratized computing, all of which support improvements in community resilience if put to the task. In a spatial context, multitudes of data from complex ecosystems and social systems can now be deciphered and made sense of in ways that simply weren't possible a decade ago.

At the time of this writing, we are also six months into a global crisis that continues to test our communities around the world in new ways. Coronavirus disease 2019 (COVID-19) has impacted 188 countries and caused more than 37 million infections and more than one million deaths since December 2019. This pandemic has led to a global withdrawal from ordinary social and cultural activities of daily life. There have been unprecedented levels of economic hardship, unemployment, government aid, and individual and community stress and frustration. We've seen some community health systems overwhelmed; hoarding mentalities that led to shortages in resources like toilet paper; and difficulty in standing up new resources, such as virus testing locations, as quickly as they are needed. In light of the turbulence of the time, GIS has never been more important as we consider questions of current and future resilience.

As the COVID-19 crisis evolves, many anticipate changes in the ways we live our public lives—in both the short term and the long term. Our communities will need to respond to these changes to provide the ongoing safety and security we'll need to stay well.

People are just now starting to take part in social and cultural activities again, and GIS tools can assist in the planning and managing of physical distancing so new transmission chains are not initiated. This will be particularly important for things like large events, entertainment venues, and even how we vote. If cases begin to increase (discovered through GIS for disease surveillance), geospatial tools can be useful in monitoring human mobility data, looking for areas of concern among ano-nymized data points from mobile devices. All types of government and health ser-vices will need map-based dashboards that capture current service demand as well as spatial models that forecast future demand so that actions can be taken to augment capacity when and where it is needed. And GIS offers an important tool to connect residents to the services they need, during blue skies and during crises. Map-based directories offer simple and intuitive ways to ensure that people can locate their nearest food distribution site, grocery store, or essential business when they need it.

In the ideal community, no one would be more vulnerable than anyone else. But the fact is that every emergency will impact members of the community unequally. GIS can improve vulnerabilities through analysis of risk and disparities, following up with adjustments and mitigation practices to provide equitable care and services for all.

The Steinbergs have noted that social change follows environmental change. We're about to see that play out globally over the next several months and probably years. COVID-19 is likely one of those once-in-a-decade events that will fundamen-tally change everything. Let's look at some other noteworthy examples.

The 9/11 terrorist attack on the United States in 2001 had impacts that rippled around the world and created an entirely new era of regulations, opportunities, and mindsets about security. We developed a more integrated intelligence system that could connect the geospatial dots across the intelligence community. The infor-mation infrastructure was improved and critical physical infrastructure gained stronger protections.

In 2005, Hurricane Katrina devastated the city of New Orleans and the sur-rounding area. It was one of the costliest tropical cyclones on record, causing $125 billion in damage and the loss of more than 1,200 lives. Through this disaster, it became clear that better national preparedness and communication were sorely needed, and we saw increased funding to support those efforts.

In a disaster commonly known as the Camp Fire, the community of Paradise, California, burned in November 2019, with the fire consuming some 14,000 homes and 4,800 businesses, schools, and churches, and killing 87 people. Through this horrible disaster, it became clear that communities needed better coordination and communication, which failed when 17 cell phone towers became inoperable. It was also clear that funding for fire preparedness and development of resilience measures such as secondary evacuation routes were sorely needed.

Resilient communities learn from past stressors. From COVID-19 we're learning that we're never as prepared as we believe ourselves to be. We did not accurately assess the risks, and despite our interconnected world, our local, state, national, and international systems and infrastructure are not connected and coordinated. It's time to pay attention so that we can better respond to the next event. As the Steinbergs point out, "Failure to do so results in loss of community, tradition, culture, economic viability, and often life itself."

I urge you to read this book with intention. Examine the patterns and processes and put them to use, making your own community more resilient.

Este Geraghty, MD, MS, MPH, CPH, GISP
Chief Medical Officer, Esri
Sacramento, California
June 2020

Preface

For over half a century, the development and applications of geographic information systems (GIS) have continually evolved and expanded. Since the turn of the millennium, spatial analysis and GIS have emerged from the domain of the specialist into a prevalent tool used daily by professionals across almost every discipline, as well as by the public, although they may not recognize the app on their phone as GIS.

Resilience is a term that has come into common use in recent years, and while it has multiple definitions, the fundamental concept is the ability of a system to respond to and recover from an event that causes a negative impact. A system could be described as an ecosystem, a community, or a culture located in particular geography. Our own interest in the concept of resilience initially came from our observations of various severe weather events and natural disasters around the world. Clearly, there were good examples of, and intersection of, GIS and the planning and response to such often unexpected events, especially in this age of highly connected communities.

As we considered our own work across a variety of disciplines, and more importantly, interacted with colleagues in both the academic and practitioner communities, we began to see valuable examples of GIS being applied to resilience in new ways, not only in contexts of extreme weather and disasters, but also in social contexts, such as in preserving traditional ways of living, recording and preserving history, and planning for the resilience of these social and community systems.

While these forms of resilience are very different, they had several common features. All faced a variety of known or unknown external stressors with the potential to threaten the long-term survival and function of the system or community, and these stressors included both human-caused and natural forces. Each example provided an opportunity to plan for and respond to these stressors in ways that might help eliminate, minimize, and respond to them over time. And finally, because all of them occur in and rely on a critical spatial context—where things were, are, or will be in the future—there is a natural opportunity for GIS to play a role in addressing the concerns of the community, managers, or others involved in assessing and addressing the issues. Being resilient is about being adaptable and being able to quickly and effectively respond to change. GIS can play an important role in achieving this.

With this in mind, we sought to identify people working with GIS in the context of resilience as a means to provide real-world examples of how this continually evolving technology can serve our larger objectives in the context of a rapidly changing world. The world has grown much smaller in the past decade, as more and more of our planet's inhabitants have gained access to geospatial tools in the palm of their hand, and not only in the industrialized parts of the world, but across the vast majority of the planet. Almost everyone these days has a cell phone, which serves as a minicomputer and geospatial locating device. These trends will only continue well into the future.

Ubiquitous technology provides incredible opportunities, and in the hands of local communities and citizens, these tools can offer new ways to capture, record, and analyze information using methods previously accessible to only a privileged few. Geospatial technology is more readily accessible and serves as a democratizing technology around the world. We found the contributions of all our authors inspiring, and, in many cases, they presented applications of GIS and resilience that we had not previously considered. Additional information about these amazing individuals can be found in the appendix. We hope you will find similar inspiration and new ideas in the chapters of this book and use this amazing technology to support resilience in your own communities.

Steven J. Steinberg, PhD, GISP
Sheila Lakshmi Steinberg, PhD
Irvine, California
May 2018

→ CHAPTER 1

Conceptualizing spatial resilience

**SHEILA LAKSHMI STEINBERG AND
STEVEN J. STEINBERG**

INTRODUCTION

How do populations react when they encounter unexpected changing environmental conditions? People around the globe frequently experience unexpected situations. Consider the coronavirus disease 2019 (COVID-19) pandemic that emerged in late 2019 and rapidly affected the entire world, while at the local level differentially impacting communities and limiting mobility across various environments. Spatial resilience is built around the keen ability for people to adapt to changes in a positive manner. For example, when a community experiences an extreme winter, heat wave, or unprecedented wildfires or storm surge, it may quickly raise an alarm. When physical environments are constantly changing, communities face a variety of challenges and stress. However, recognizing and responding to such changes and taking action to positively respond can result in spatial resilience.

The earth is a dynamic planet where physical environmental norms are in flux daily. As the physical environment around us begins to change, the social and environmental experiences of individuals fluctuate, resulting in changing places and spaces. One day you may be living in a community by the coast, and the next day your home is washed away by a tsunami. Powerful storms occur at times of year when they are not typically anticipated, which can have disastrous results for people and place. Erosion occurs, lives are lost, and settlement patterns experience disruption.

Large regions of the globe are experiencing major drought and fires, while others are overwhelmed by excessive precipitation, landslides, or coastal flooding.

Almost daily, extreme weather happens, with many record-setting statistics. Regardless of the underlying cause, alternations in weather, climate, and physical environment represent a change from what has been considered normal to create a new reality. When environmental change occurs, social change naturally follows. Such environmental and social changes demand a sociospatial response. For instance, if you live in an area that experiences storm surge now more than in the past and your home floods (or, for example, if you live in the city of Venice, Italy), you may need to move to higher ground. Or in the case of Venice, perhaps the city needs to be raised or flood control mechanisms need to be installed. Spatial resilience is a skill that populations who live in such areas must develop as part of their survival and adaptation to changing environmental conditions.

ENVIRONMENTAL CHANGES

Environmental changes are visible across a variety of geographies: mountains, deserts, coastal areas, and urban and rural environments. Shifting weather patterns create major changes in the environment, presenting numerous challenges. For communities that experience these environmental changes, there is often no clear prescription or plan for how to react. This book provides a foundation for individuals, communities, governments, and agencies to explore such changes by harnessing spatial data and analysis approaches as they plan for long-term resilience of their communities and environments.

Human communities organize themselves around an established sense of our destiny and our future. When unexpected changes interrupt our plans, it is not a welcome change— why? Because addressing these major environmental and social interruptions requires change on our part—often change for which we do not have the resources to enact. This is especially true when these unexpected changes require major alterations to the way some people lead their everyday lives. But the reality is that environmental change is happening more frequently and extremely today than ever before, and it is wreaking havoc on many geographies. The question is, how are communities going to cope with this across different geographies?

We chose to write this book focused on the application of geographic information system (GIS) technology for spatial resiliency analysis and planning because we recognize that traditional physical environments (ocean, coastal, mountain, desert, urban, and rural) are experiencing previously unknown and unprecedented environmental changes. The socioenvironmental changes are coming quickly, unexpectedly, and with great intensity. Events over the past several years have spurred us to think in an interdisciplinary fashion (as a team of physical and social scientists) about these changes and to consider what might be done to be better prepared and, ultimately, more resilient.

Additionally, the changing nature of ethnicity in certain environments needs to be considered. So often the names associated with place are determined by communities of a historic origin. When the people who populate a certain area change, the place-names may stay or may also be altered. Perhaps no group is more original for a particular location than its Indigenous peoples. Throughout our book, we focus on Indigenous populations and their interaction with place and space. GIS is examined and used as an important tool for helping Indigenous communities maintain resilience in the face of changing physical and structural environments. We focus on how GIS can maximize Indigenous groups' connection to and use of the resources in their environments.

In addition to Indigenous groups, we focus on the important role that ethnicity plays in mediating other cultural groups, such as Latino interaction with surrounding natural environments and communities. The intricate layers of interaction and community can vary across places, often depends on the cultural norms and practices of the people who live there. Some might say that the surrounding physical environment impacts local social interaction in a particular place. We would say that interaction can indeed be impacted by physical environmental structures but is mediated by the cultural background of a particular group.

We also considered questions such as, How do we get a good sense of a problem or situation? How do we process information? How do we gather data? What data should we consider? And ultimately, how do we assess the current patterns and changes in our surrounding environments that are most important? These represent questions that we explore as the examples in this book. Using information and analysis tools more effectively, we hope communities experiencing these changes will be better able to respond to, react, and plan for future resilience. Failure to do so results in loss of community, tradition, culture, economic viability, and often life itself.

Being prepared to be resilient requires communities to first understand who they are in the context of their own environment and then to thoughtfully prepare for environmental change. Through this book, we provide a methodological framework, case studies, and lessons learned about how such planning and assessment can be effectively applied using the capabilities of GIS as a key tool in the process.

In our own research, we've found it is best to observe a situation, problem, or issue from multiple perspectives, considering both physical and social environmental factors and the characteristics of the place that affect its resilience. Resilience is the ability to manage the shifts between environment and society in a manner that produces balance and harmony. It's of the utmost importance to begin this journey by understanding people in the context of their place and to harness local knowledge, skills, and abilities for how they interact daily with their environment. Using this information as a starting point is necessary before one considers response and action to environmental changes.

RESPONDING TO VULNERABILITY

The weather, climate, and overall physical environments are in a constant state of flux, perhaps now more than ever. This process creates vulnerability. Weather records are being broken continually, and people are frustrated because much of what they know about the places and environments they inhabit no longer holds true. The ways of life that many communities know and have experience with—including patterns of interaction that they have established over time with their local places—are being severely challenged. Inevitably, such changes create stress through exposing the vulnerabilities of people and places. People become more vulnerable when they live in a city or community that is under the stress of changing environmental conditions that are unfamiliar and unexpected. It becomes an issue of magnitude and frequency.

For instance, if you live in one of the urban cities on the East Coast of the United States, such as Boston, you expect there to be snow, but not two to three feet of snow from one storm. You expect to have big blizzards and storms every now and then, but not the kind of megastorms that have occurred over the last few years. Normal patterns of mobility and action are halted by these major environmental changes, such as unexpected intense snowstorms. The result? People suffer and can't get to work, can't get to the hospital, and are generally snowbound.

Extreme weather and climatological changes are occurring now on an order of magnitude and with increased frequency and severity that demand attention. This increase demands attention because when major shifts in the physical environment occur, the social community and patterns of interaction shift as well. Reviews of scientific data show that there is more carbon dioxide in the atmosphere now than at any other time in the last 400,000 years (NASA 2015) and that temperatures today are hitting the highest levels ever recorded. Additionally, issues such as drought or water shortages continue to impact societies in the United States and abroad. Societies are not just sitting by and waiting for disaster to strike; they are assessing the situation and coming up with a plan of action.

Many of the physical environmental changes that are occurring these days are extreme. What is a community to do if the sea level rises and destroys half the town? What if there are severe mudslides due to too much rain and deforestation in a remote mountain community? How should the local residents and natural resource professionals react to increasing mudslides, resulting in increased physical and economic isolation? Another example might be desert communities, where water is the main limiting factor. How about the large cities and places that are supposed to thrive in increased drought? There are only so many resources available, especially when it comes to water, so where do we find these resources, and how do we use them to sustain communities that are in challenging geographies? These are all major questions that illustrate some of the physical environmental shifts that communities increasingly face.

But the question remains: With all the data that is floating around, what is the best way to be ready to respond to changes in a productive manner? This can be a difficult question to answer, but one can begin by adopting a theoretical approach as to how to conceptualize people in the context of such environmental changes.

Figure 1.1 presents the spatial resilience model to illustrate the major factors that should be considered as we think about people, environmental change, and how communities are impacted. The model consists of four concepts: environmental change, society/community, action/policy, and resilience. In the model, you see that we begin with environmental change, which impacts society/community, and similarly, society/community also impacts and influences environmental change. This two-way flow highlights the reciprocal relationship that exists between these concepts. The point of this analysis is to result in action and policy. The orange box at the top of the model represents resilience. Resilience is the overarching component of the model that positively impacts the two-way flow relationship between environmental change and society/community. This implies that those communities or places that are resilient will be able to better weather and respond to the physical environmental shifts and changes than those communities that are less resilient. Furthermore, there will be a direct relationship between spatial resiliency analysis and action policy. Why?

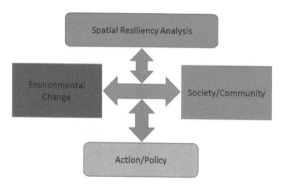

Figure 1.1. A model for spatial resilience.

In essence, the heart of the model is in the spatial resiliency analysis. If communities are able to know, understand, and react to or take charge of their destiny through analyzing their environments, they are on the path toward being resilient. Our approach is based on the idea that the most effective action/policy will derive from the environment where people know their strengths, understand their weaknesses, and therefore geographically have a better sense of where to target their action and efforts. That is what spatial resiliency analysis is all about. Throughout this book, we will explore the various aspects and features of spatial resiliency analysis using GIS.

Societies and communities that successfully adapt to changing conditions will grow and prosper. For example, the Ancient Pueblo people (often known as the Anasazi) were once a very sophisticated society centered on the four corners area of New Mexico, Arizona, Colorado, and Utah. However, around AD 1200, their society disappeared (Roberts 2003). No one really knows exactly what happened, but some researchers have postulated that changing access to food, perhaps due to drought and changing climatological conditions, may have played a role in their disappearance. Additionally, other scientists have argued that their disappearance was due to warfare and political encroachment by other tribes. In any case, the lesson to be learned from the Anasazi societal disappearance example is that there can be societies that are very well and clearly established in a particular geography, and then something occurs to change that. In fact, when we witness the disappearance of a society, clearly a particular vulnerability of that society—whether it be environmental, social, or political—occurred.

The ability of a people to respond to change, especially when it threatens their vulnerabilities or weaknesses, is especially important. In this book, we explore how effective data analysis, planning, and assessment have led to smart policy action on the part of leaders and decision makers to bolster the strengths of the community. In other words, we explore how societies can be truly resilient and become active in their response to change versus being vulnerable and negatively impacted by these changes.

There are many such examples throughout history that highlight the importance that being adaptive or resilient to change can play. One could say that the Anasazi people faded out or disappeared because they lacked resilience (or the ability to productively adapt and successfully bounce back) to the physical environmental changes that occurred around them. They were not adaptive enough as a society to successfully make their way through the changes that occurred.

GIS AND SPATIAL VIEWPOINT

The idea for this book emerged from our experience with spatial analysis and particularly GIS. GIS is a computerized mapping technology that has a great capacity to be used in data capture, analysis, and visualization. For many years, GIS has been an effective tool used in the military, planning, and natural resources fields to take stock of resources and situations and prepare for action. Today, GIS has made its way into almost any field that you can think of: health, medicine, sociology, planning, natural resources, business, and marketing, just to name a few. The application of spatial thinking around the globe has burgeoned because people realize the usefulness of thinking spatially.

GIS allows people to actually see data, sometimes even in three-dimensional formats. When you can see data as a readily accessible information product that makes the information visual, it can help to enable a better understanding and comprehension of patterns. But more than just allowing you to visualize data, GIS is a highly sophisticated tool for data analysis and integration. GIS is a means for groups to capture the story of people in a particular place. It enables the charting of environmental change and connecting it to sociodemographic change. When a major hurricane is about to strike, who has the ability to pick up and to move out of its path? GIS enables one to put multiple layers together to look for patterns in data. Through assessment of these patterns, we begin to understand how people engage with place along with unique coping strategies and reactions to change that we can term *resilience*. Assessment and understanding of a problem or issue will be much better if that process is triangulated—meaning measured from various angles and not just one perspective. When one uses GIS, one can clearly see resources or lack of resources.

This capability is especially useful for leaders of countries, the military, city managers, and policy makers to be able to know and understand the lay of the land. In other words, GIS can simultaneously provide a bird's-eye view of a situation or problem along with a more specific analysis of a particular geographic location or area.

RESILIENCE

Resilience means being able to account for the change that occurs between physical environmental change and societal/community change patterns in a positive and proactive manner. Resilience is not only an ideal but also something that people and communities can actively strive to reach through careful planning and assessment. The thesis underlying this book is that if people have solid data about their communities, they can use it to spatially understand the strengths and weaknesses of their local geographies and the various environmental challenges that exist.

The spatial resilience process means taking stock of what you have, trying to target and understand vulnerabilities, and then moving forward to ready your community to account for change. If you successfully read the data, you can identify weaknesses and areas that need bolstering before disaster strikes. If you are aware of past context and history of environmental changes, including how people responded to changes, you can successfully respond to these changes in the future. A major part of achieving resilience is understanding people in the context of their geographic place.

PEOPLE IN THE CONTEXT OF PLACE

People are always tied to a particular place or geography. As they establish themselves in a particular physical geography and climate, they establish ways of interacting with that environment. A community's patterns of interaction with its environment are going to be influenced by a number of things in the physical environment, such as weather, stability of the land, nature of the topography, and proximity of the geographic area. These factors will in turn be mediated by social factors, including politics, culture, social structure, and power. For instance, major environmental shifts could result in further isolation for a particular place or could change the season of an entire industry (such as agriculture) that is dependent on certain climatic conditions to achieve production goals.

Given all the environmental shifts that are happening both locally and globally, as scientists and authors, we wanted to share information about how using the spatial perspective can lead to successful resilience. The responses will be holistic in nature because spatial analysis gives a larger, more realistic view of major environmental changes that occur in various geographies. The best way to establish useful plans for action and response to change is to first gain a solid understanding of your environment through gathering baseline information about physical geography using various types of data; analyzing that data to identify strengths, weaknesses, and patterns; and developing action plans based on the analysis. This book focuses on change and the ability to successfully plan for and adapt to change in a manner that works or is successful. Resilience is defined as an ability to recover from or adjust easily to misfortune or change (Merriam-Webster, n.d.).

INTERDISCIPLINARY PERSPECTIVE

This book is unique because we incorporate an interdisciplinary approach to looking at the issues of resilience. The fields that are covered in this book include geography, social sciences, planning, landscape architecture, urban and rural sociology, economics, migration, community development, meteorology, and oceanography. But the main distinctive feature of our book is that we encourage, recognize, and highlight the integration of various types of data (quantitative, qualitative, and spatial) to produce a holistic view of a challenge. We encourage communities to achieve resilience through an integrated, thoughtful, spatial approach. That is where GIS technology emerges as a leader in the field of spatial resilience, because it enables this data integration and the consideration of various pieces of the story that create a holistic view.

This book is a hybrid of theory and action, consisting of background and practical steps for how to achieve resilience. Each chapter was contributed by experts who have used GIS as part of their work on resilience in their respective fields. This book is structured to provide the reader with both theory and applied examples supporting and illustrating this theory. We hope to take you on a journey through a variety of environments and applications to illustrate resilience in context of a particular environment and community that incorporates a spatial perspective and uses GIS technology in the approach. While GIS is not always the primary focus of the cases presented, in each of these examples it provides essential tools and capabilities necessary to assess, explore, and develop policies or solutions that address aspects of resilience in the particular case.

WHY RESILIENT COMMUNITIES?

We selected the case studies presented in this volume to highlight a variety of ways resilience, and the analysis of resilience using GIS, can be applied across multiple environments and communities. A theme carried throughout these narratives is that community members, decision makers, and researchers working across varied geographies can effectively use spatial analysis as a component of their efforts and plan to adapt to changing natural and cultural environments in a productive manner to produce more resilient communities. Regardless of the underlying causes, change will happen in spaces and places where people interact with their environment. Sometimes these changes are rapid or unanticipated, while other times changes may occur in the long term. Some may be influenced by the very changes made by humans as they alter the physical and natural environments to serve the needs of their changing world, communities, and economies. These changes may be subtle or fairly significant due to an alteration in a physical ecosystem. Communities interact with their environments in many ways, fine-tuning these spaces and places to adjust to or to influence changing relationships and patterns.

Communities around the globe work toward resilience on a regular basis, and a goal of resilience—or more simply, an ability to sustain a community and its standard of living—is something that is common to all communities. All societies seek to maintain and enhance their environment, in whatever manner this is defined in their own culture. Everybody in an existing locality can be involved in contributing key data essential to understanding people and their context in a place. A strong community arises from the local assets that exist in a given geography. Throughout this book, we share stories of resilience and the successful navigation of challenges that emerge across a variety of contexts.

Living and working in a community necessitate navigation around the local spaces and places, being aware of anything that is happening in the local ecology. The awareness and familiarity that local people have with a place too often go unnoticed and unrecognized. What we have highlighted here is that people who care about a locality often possess the will and skill to engage in sociospatial ecological thinking. This means that communities can situate themselves into larger contexts or surroundings. They have the keen ability to visualize the environment and to see their position, role, or place in the larger ecosystem. GIS can play an important role in that process, as we see in the various spatial case studies presented in this book. GIS is a powerful tool because it facilitates the overlaying and visualization of different types of data. When you can see data portrayed, it makes it easier to see the patterns of where certain features occur.

For any society, the ability to manage and adjust to changes in the physical, social, economic, and political environments is powerful. Throughout history, societies have consistently encountered changing surroundings in their social and physical environments. For instance, there is the mystery of the Anasazi, who had well-developed and complex societies and seemingly disappeared without a clear explanation. According to science journalist George Johnson in *The New York Times*, "Like people today, the Anasazi (or Ancient Puebloans, as they are increasingly called) were presumably complex beings with the ability to make decisions, good and bad, about how to react to a changing environment. They were not pawns but players in the game" (Johnson 2008, 1). Some scientists hypothesize that the seeming disappearance of the Anasazi was due to drought or other changes in the climate. Other explanations suggest the onset of a plague or sickness. The bottom line is that when societies fail to adapt to major environmental or social changes, they suffer the consequences of these changed ecosystems or social constructs in their world. As a result, many societies fade away, sometimes for well-understood reasons, or, in other cases, without any clear explanation.

People live in various neighborhoods, districts, villages, and towns for reasons that can be teased out of the data. These reasons may relate to the availability (or historical availability) of specific resources in the surrounding physical environment, including natural resources essential to the function of the society or secondary values such as access to transportation or trade routes. Sometimes societies develop in environments that are less hospitable, often the result of social or political influences that force a population into places they might not otherwise choose, for example, the forced resettlement of Native Americans to reservations.

Patterns of local social interaction and engagement are often driven by the physical geography of where people live. This environmentally driven, site-specific social interaction is not something that we discuss enough. However, when people have meaningful knowledge about local places and geographies that can be included in the data collection and mapping process (for example, by incorporating public participation GIS methods), communities have a better chance to achieve resilience in part by maintaining control of their own local knowledge.

SPATIAL ECOLOGY OF PEOPLE AND PLACES

What is spatial ecology? It is the established social and spatial connections between people, space, and place in an ecosystem. As we think about people and place, a valuable lens through which to view change and the resulting adaptive responses is sociospatial ecology. Sociospatial ecology means holistically considering the spatial and environmental connections between and among entities in a bounded area.

Tobler's First Law of Geography states, "Everything is related to everything else. But near things are more related than distant things" (Tobler 1970). According to this line of thought, entities that exist proximate to one another will have more in common with one another than those that are farther away. In other words, the objects within a physical boundary are going to be more in line with one another than those not located within the same physical boundary. Perhaps less known is Tobler's Second Law, stating, "The phenomenon external to an area of interest affects what goes on inside" (Tobler 1998). In other words, while the factors within an area, perhaps within a community, possess important internal relationships, these things do not exist in a vacuum. People and places are also affected by factors external to them, such as natural, social, or political events.

Communities are really miniature ecologies that exist at a specific time, space, and place, usually bounded by some sort of loosely defined border. The border serves as an edge or limit on the system, but entities also pass from outside the system to inside the system and back again. Spatial ecology focuses on spatial patterned relationships between entities within a bounded ecological system, allowing you to look at things internal to a community as well as the ebb and flow of physical resources and human and social capital resources in and out of an ecosystem.

SPATIAL CASE STUDIES

The case studies we compiled for this book exist across a wide variety of physical spaces. We grouped these into four thematic sections to help guide readers to topics that are most relevant to their interests. Nonetheless, each of these case studies includes two common themes: first, an effort or need to adapt to changes in the natural or social environment, and second, a use of GIS technology to help facilitate the collection, visualization, and analysis of this information. Regardless of your own interests, all of the examples presented incorporate multiple methods and interesting approaches, so we encourage you to explore examples from each section. Many of the methods used to explore these issues of resilience may be transferable to your own area of interest, even if the specific example comes from a different geography from your own.

Given the right tools and data, any community or organization has an opportunity to solve its problems through critical spatial thinking and analysis and, ultimately, is better equipped to make decisions. Of course, communities have managed for resilience long before the development of GIS. Spatial thinking doesn't require technology. However, GIS provides a powerful enabling technology to effectively assess, consider, and communicate options in ways that were not easily accomplished before its availability. GIS enables people, place, and space to interact together as a collective whole to respond to change in an informed, targeted manner versus relying on an incomplete, qualitative awareness of changes here or there in a particular environment. GIS can help communities prepare for change and adapt to changes as they occur in real time. Furthermore, the analytical capabilities of GIS offer valuable predictive capabilities that can be harnessed to help communities achieve the best results possible.

It goes without saying that in most cases, the members of a local community will have the greatest understanding and awareness of their surroundings based on the learned and accumulated knowledge that develops and is passed down over time. As such, local populations can be quick to recognize and respond to changes that occur, especially when these changes relate to core elements of their survival, food, shelter, and the safety of their population. The result is that communities act when needed and adjust themselves.

CLIMATE CHANGE AND WATER GEOGRAPHIES

Resilience in ocean and coastal areas is highlighted in chapter 2, "Resilience in coastal regions: The case of Georgia, USA," by Rosanna G. Rivero, Alison L. Smith, and Mariana B. Alfonso Fragomeni. The authors explore various factors that impact coastal vulnerability to climate change through adopting a multidisciplinary approach to achieving coastal resilience in Georgia and also explore the use of a unique geodesign framework.

A second water-focused chapter concentrates on safety and access to the Columbia River. Written by Paul Cedfeldt, Jacob Watts, Hans Moritz, and Heidi Moritz, chapter 4, "The mouth of the Columbia River: USACE, GIS and resilience in a dynamic coastal system," tells the story of how GIS is used to achieve resilience at the mouth of the Columbia River and highlights the history of the river, its people, and how the United States Army Corps of Engineers works with the community to help achieve greater resilience.

INDIGENOUS ECOLOGIES

We have included two chapters examining the use of GIS by Indigenous and minority communities. These chapters demonstrate how gathering local knowledge of a community and its environments and packaging it using GIS can better preserve, document, and communicate this important cultural knowledge. These chapters highlight the power of Indigenous knowledge and the role that GIS can play.

In chapter 7, "Indigenous Martu knowledge: Mapping place through song and story" by Sue Davenport and Peter Johnson, we see how one Australian Aboriginal community, the Martu people, has worked to rebuild social stability and resilience. The key is a combination of ancient knowledge and practices with modern technologies. This population often leads a traditional desert existence and uses songs and stories to share information and to cope with the modern surrounding physical environment. GIS provides a means to capture and use this oral tradition in new ways.

Authors Kevin O'Connor and Bob Sharp wrote chapter 8, "Developing resiliency through place-based activities in Canada." Their chapter explores how place-based education in a rural, Indigenous community empowered local youth to make a difference in understanding community history, culture, and local traditional ecological knowledge. Students developed a better understanding of themselves situated within the larger community and natural world. The chapter examines the connection to social and spatial place. When students can contribute to the resilience of their community while using new and exciting tools such as GIS, it both empowers them to make a difference in their local communities and offers a new set of knowledge and skills that will benefit their own lives. Helping their community to capture and use local knowledge gives them the power to act to benefit the entire community.

URBAN ECOLOGIES

Author Jason Douglas focuses on youth in an urban environment in chapter 9, "Engaging youth in spatial modes of thought toward social and environmental resilience." This chapter highlights how an urban mapping program empowers the community to focus on resilience and various environmental justice issues. Public participation GIS is employed, and students are again the leaders, working with their communities to enable change.

A second urban example comes from chapter 5, contributed by Regan Maas. "Urban resilience: Neighborhood spatial complexity and the importance of social connectivity" explores how poorer communities have a great strength in terms of their social networks. GIS is used to examine cluster analysis and how to identify resilient neighborhoods and individuals. Populations that face adversity can ultimately develop the skill sets necessary to come out on top by becoming adept at adjusting to and fighting their way through challenges.

Chapter 3, "Building resilient regions: Spatial analysis as a tool for ecosystem-based climate adaptation" by Laurel Hunt, Michele Romolini, and Eric Strauss, focuses on creating resilient cities and regions across five similar Mediterranean-climate regions: California, Central Chile, the Western Cape of South Africa, South and South West Australia, and the region bordering the Mediterranean Sea. Despite different geopolitical boundaries, comparisons are drawn across similar Mediterranean ecologies.

RURAL ECOLOGIES

In chapter 10, "Health, place, and space: Public participation GIS for rural community power," which we contributed, we examine how GIS can be used for community empowerment in rural Latino communities. We worked with community-based organizations and local residents to examine, document, and address issues related to agricultural pesticides and community health, ultimately leading to new policies for buffer zones preventing the spraying of pesticides near schools. This chapter ties together a number of sociospatial methods, including public participation GIS, use of pesticide application permit data from local government, and information on local weather patterns and crops to assess and communicate information to the local communities. Our analysis provided the basis for developing policies that support the needs of both the local growers and farming communities in California, which depend on each other to support the economy of these regions and provide fruits and vegetables throughout the country and beyond.

AS YOU EMBARK ON THIS JOURNEY

We hope you will find the examples embedded in these spatial case studies as a source of inspiration for your own work. Please use and modify the methods and best practices presented in these examples to achieve spatial resilience and effective policy in the communities and geographies where you work. Successful policies to achieve long-term resilience are those that actively consider the strengths of people in the context of their geographies.

REFERENCES

Johnson, George. 2008. "Vanished: A Pueblo Mystery." *The New York Times*, April 8, 2008.

Merriam-Webster. n.d. "Resilience." In *Merriam-Webster.com*. Accessed May 24, 2020. www.merriam-webster.com/dictionary/resilience.

NASA. 2015. "The Relentless Rise of Carbon Dioxide." NASA Global Climate Change: Vital Signs of the Planet. Accessed May 23, 2020. https://climate.nasa.gov/climate_resources/24/graphic-the-relentless-rise-of-carbon-dioxide.

Roberts, D. 2003 "Riddles of the Anasazi: What awful event forced the Anasazi to flee their homeland never to return?" *Smithsonian Magazine*, July 2003.

Tobler, W. R. 1970. A Computer Movie Simulating Urban Growth in the Detroit Region, *Economic Geography* 46 (Supplement): 234–40.

Tobler, W. R. 1998. Linear pycnophylactic reallocation—Comment on a paper by D. Martin. *International Journal of Geographical Information Science* 13 (1): 85–90.

→ CHAPTER 2

Resilience in coastal regions: The case of Georgia, USA

**ROSANNA G. RIVERO, ALISON L. SMITH, AND
MARIANA B. ALFONSO FRAGOMENI**

INTRODUCTION

In 2010, the United States Census Bureau estimated that 39 percent of the United States' total population lived in coastal shoreline counties. Current trends indicate an expected 8 percent increase in population in coastal regions between 2010 and 2020 (NOAA 2013). In conjunction with this expected population growth, coastal areas are particularly vulnerable to climate change, including effects of sea level rise and flooding, which are gaining public attention due to the observable nature of such occurrences. Furthermore, changes being made to the United States' federal flood insurance (Biggert-Waters 2012; US Congress, House 2014) have affected coastal communities and demanded plans that look closely at the issues of storm water management, which are linked to flooding and sea level rise. The combination of a growing population with an evident vulnerability to water-related events makes the coast an opportune location to apply a geodesign framework. The use of this method allows for a multidisciplinary approach that recognizes and considers the complexity of environments, such as the coast, where socioenvironmental processes occur simultaneously.

In the southeastern United States, the impact of major flooding events and natural disasters along with the response by the affected communities is not a new topic of interest. Between 1926 and 1928, two major hurricanes struck in Florida, resulting in over 2,000 deaths (Gannon 1993; Grunwald 2007). According to the National Hurricane Center, the Okeechobee hurricane of 1928 was officially classified, by 2003, as the second-deadliest disaster in American history, behind only the Galveston hurricane of 1900, which killed between 6,000 and 12,000 people, and the Johnstown, Pennsylvania, flood of 1889, which killed 2,209 (Kleinberg 2004). The consequence of this major event was the construction of the Herbert Hoover Dike around Lake Okeechobee, which has generated other environmental consequences for the restoration of the South Florida coastal regions and the Everglades (Grunwald 2007).

Other damaging hurricanes have followed in the southeastern United States throughout the years: Andrew (category 5 in 1992), with damages estimated at $26 billion and a loss of 25,524 homes (Rappaport 1993), and Katrina (category 3 in 2005), with about 1,200 fatalities. Hurricane Katrina is considered the third-deadliest hurricane in the United States, after Galveston (1900) and Okeechobee (1928), and also the costliest, with a total estimate of $108 billion in property damages, roughly four times the damage created by Hurricane Andrew in 1992 (Blake et al. 2011). A recent report from NOAA (2020), with adjusted cost values based on the 2020 Consumer Price Index, shows the following rankings after incorporating recent events: Hurricane Katrina (2005, $170.0 billion) ranks first; Hurricane Harvey (2017, $131.3 billion) ranks second; Hurricane Maria (2017, $94.5 billion) ranks third; Hurricane Sandy (2012, $74.8 billion) ranks fourth; and Hurricane Irma (2017, $52.5 billion) ranks fifth.

Natural disasters such as Superstorm Sandy in New Jersey (2012) point to the added threat of sea level rise and the potential for historical flooding in areas perceived as safe or at low risk. In fact, hurricanes and tropical storms do not need to make landfall to threaten coastal communities. In October 2015 in South Carolina, a stalled front offshore combined with deep tropical moisture, an upper-level low-pressure system, and Hurricane Joaquin in the Atlantic Ocean to result in historic flooding along the state's coast (NWS 2015).

By contrast, along the coast of Georgia, unlike other states in the Southeast, there was a perceived notion of safety, given that no major hurricanes had impacted the state's shorelines since the late 1800s (Blake et al. 2011). That was until Hurricane Michael (2018), which is now considered one of the top 10 costliest hurricanes in the US. Communities in Georgia are not free from problems related to the combined effects of sea level rise, storms, and hurricanes. For instance, cities such as Tybee Island and St. Mary's have sought help from institutions such as Georgia Sea

Grant and the Carl Vinson Institute of Government, at the University of Georgia, to address flooding and storm water management. Scenarios such as king tides have become more frequent along US Highway 80 (figure 2.1), the only highway between Tybee Island and the city of Savannah, currently flooding four to five times per year. Such an occurrence, combined with increased sea level rise, has made flooding on Tybee Island more frequent, and it is expected to increase if protective measures are not taken (Evans et al. 2013). As a result, decision-makers and residents are seeking to address the problem through the development of long-term plans that could enable the city to become resilient to flood events.

Figure 2.1. View of Tybee Island after Hurricane Matthew.

Photo by Paul Wolff.

The issue of development versus conservation of coastal environments is important in the context of resiliency. In this sense, compared to other states in the southeastern United States, Georgia is unique, given that the state has almost 100 miles of relatively pristine and undeveloped coast. In contrast, its neighboring state, Florida, has approximately 1,350 miles of coast and 15.9 percent of the United States coastal population (US Census Bureau 2010). Georgia has only 0.5 percent of the coastal population, followed by South Carolina and North Carolina holding 1 percent each (figure 2.2). With the addition of natural hazards that have occurred on the East Coast in the last 100 years—particularly in recent times, such as Superstorm Sandy in 2012, the historic floods in South Carolina in 2015, and the recurrence of events such as king tides in Tybee Island (Landers 2014; Byrne 2015; Galloway 2015)—the subject of how the coast is prepared for long-term natural events becomes an important focus in any conversation about resilience. Projections demonstrate that other major changes could be expected in the next 45 to 50 years due to a combination of factors, including population growth and global climatic threats such as sea level rise, storm surges, and droughts (Rivero et al. 2015; Hauer, Evans, and Mishra 2016).

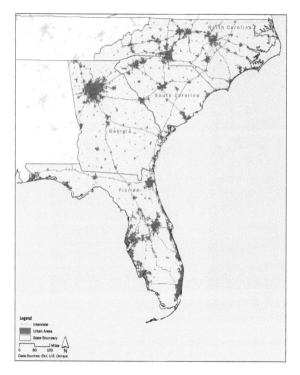

Figure 2.2. United States Eastern Seaboard showing urban areas in gray. Notice the extension of urban areas in neighboring states to Georgia.

Map by Alison L. Smith; data courtesy of Esri, US Census Bureau.

It is important to point out that the discussion of resilience in coastal regions is a global matter and has gained the attention of planners and decision makers throughout the world. Likewise, in the southeastern United States and along the coast of Georgia, this has become an important topic for planners and for local, regional, and national organizations, including counties, the Coastal Regional Commission, nongovernmental organizations such as The Nature Conservancy and the GA Conservancy, and academic institutions including the University of Georgia. These organizations specifically study the intersection of development and conservation, seeking to ensure that coastal communities can cope and bounce back from possible impacts of natural hazards and a changing climate.

Multiple plans have addressed resilience, in some cases with multiscale approaches, from regional to local levels (Louisiana Recovery Authority 2007). The cases presented in this chapter resulted from an agreement initiated in 2013 by the College of Environment and Design at the University of Georgia and the Coastal Regional Commission of Georgia. This agreement resulted in a series of planning, educational, and service activities (including studios and workshops) framed under pedagogical concepts of experiential learning and methodologies. Some of these collaborative planning and interdisciplinary design approaches have been framed under the term *geodesign* (Steinitz 2012; Rivero et al. 2015). Among these collaborations, the 2015 Coastal Georgia Geodesign project described in this chapter has been an important milestone within the geodesign community because it was the first of several collaborations between the Centre for Advanced Spatial Analysis at University College London, the Coastal Regional Commission of Georgia, and the University of Georgia, as well as the basis for many other projects.

THEORETICAL FRAMEWORK AND KEY CONCEPTS

Resilience

For the purpose of this chapter, focused on planning and coastal areas, we introduce general concepts of resilience as applied to this discipline and environment. Recent events have made the vulnerability of coastal cities evident, and planners have been challenged to develop solutions to adapt cities to rising waters and flooding, seeking to promote resilient communities. The term *resilience* in planning, as for many other disciplines, derives from the original definition used by ecologists (Holling 1973). It is a measure of how fast a system returns to a state of equilibrium after a disturbance. Bringing this definition into the planning perspective, Eraydin and Tasan-Kok (2013) identify resilience not only as a community's ability to readily respond to a disturbance but also as dependent on three central features: "(1) the ability of a system to absorb or buffer disturbances and still maintain its core attributes, (2) the ability of the system to self-organise and (3) the capacity for learning and adaptation in the context of change." The system in this case is the entire urban network, accounting for both social and physical characteristics of the environment that will collectively respond to disturbances.

The term *vulnerability* has become commonly used in planning and other disciplines as a way of defining social and environmental exposures to risk (Cutter, Boruff, and Shirley 2003; Cutter and Finch 2008; Mendes and Tavares 2009; KC, Shepherd, and Gaither 2015). According to Muller (2011), vulnerability and the ability to recover from disturbances are not merely confined to natural hazards and climate change. The concept is also closely tied to cities' abilities to recover, reorganize, and reinvent themselves when faced with changes and adversity. With this in mind, some authors call for a broader understanding of the term *resilience* as applied to urban environments, not necessarily confined to natural hazards and climate change, but instead addressing other facets of the vulnerability of urban areas. This is the case with cities such as Detroit, Cleveland, or Pittsburgh in the United States or Manchester and Liverpool in England. In Latin America, cities such as Caracas, Venezuela, are subject to other impacts: urban decline and vulnerability to economic, demographic, and even political factors. Some of this is a product of economic crisis, deindustrialization, and political changes. These cities face the challenge of recovering or providing urban services under conditions of distress or reorganizing and eventually reinventing themselves (Müller 2011). This call for a broader understanding of resilience is addressed in our chapter by offering a systemic approach to the understanding of a region or a city.

In the United States, several agencies working on resilience-related topics have been uniting efforts. For example, in 2010 the Environmental Protection Agency (EPA) and the Federal Emergency Management Agency (FEMA) signed a Memorandum of Agreement to make it easier for the two agencies to work together to help communities become safer, healthier, and more resilient (EPA 2016). The agencies collaborate to help communities hit by disasters rebuild in ways that protect the environment, create long-term economic prosperity, and enhance neighborhoods. FEMA and EPA also help communities incorporate strategies that improve quality of life and direct development away from vulnerable areas as components of their hazard mitigation plans. In 2012, EPA—in partnership with FEMA, the National Oceanic and Atmospheric Administration (NOAA), and the North Carolina

Department of Environmental Management—worked with the coastal towns of Wilmington and New Bern, North Carolina, to address issues of vulnerability and enhance resilience.

Geodesign

Geodesign is a framework to facilitate the decision-making process when envisioning, planning, and designing—in a collaborative and multidisciplinary environment—the future of a region, a city, or a local landscape. Jack Dangermond, the founder of Esri®, offers his view on geodesign by recognizing that it is not a new concept and that it simply borrows from a number of domains—architecture, engineering, landscape architecture, urban planning, and traditional sciences—to take a holistic and complementary view on the design process, incorporating the different stakeholders (Dangermond 2010).

Many authors have offered their view on geodesign, recognizing that it is more of a framework or a collection of methods and processes than a single method or approach (Steinitz 2012; Ervin 2015; Flaxman 2010; McElvany 2012; Orland 2015). However, the principles of a holistic/systemic approach to overlaying layers of thematic information, inspired by Ian McHarg, and his book *Design with Nature* (1971) and other ecological principles, have influenced these approaches. In a traditional setting, the analyses and synthesis of these layers of information (soils, geology, land use, hydrology, and so on) were done using manual overlays conducive to identifying optimal locations for allocation of various land uses and furthering the designs for the future of an area. Collaboration, in these circumstances, was conducted through series or iterations of drawings and consequent meetings conducive to reach consensus or agreement after evaluation of various alternatives. This process could take weeks or months, depending on the size and the complexity of the project.

In a geodesign approach, the collaborative process is facilitated by computer applications that respond to changes in a design as it is being built by various stakeholders. What makes geodesign fundamentally different from the traditional design process is the workflow or the process of creating a design. The ability to create a design collaboratively, to measure the impacts of this creation as you proceed, and to implement a platform of collaboration and communication form the basis of the geodesign workflow (Rivero et al. 2015).

In his book *A Framework for Geodesign: Changing Geography by Design*, Carl Steinitz (2012) introduced his framework, which starts with six questions and three iterations (figure 2.3), developed while analyzing and refining the geodesign process.

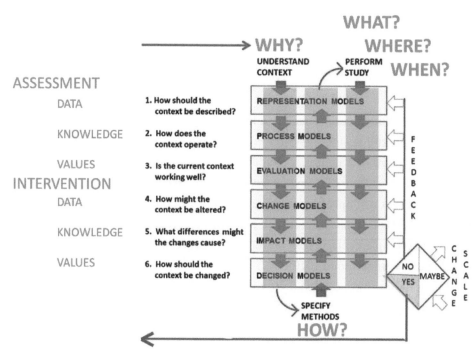

Figure 2.3. The key questions in the geodesign framework by Carl Steinitz.

Printed with permission from Carl Steinitz.

Walker et al. (2002) point out how decisions are made in social-ecological systems, in many cases with limited resources and imperfect knowledge. The idea of rapid responses, flexibility in the collaborative decision-making process, and Walker's notion of imperfect knowledge are also the basis of Steinitz's collaborative geodesign (Steinitz, personal communication, 2016). With new digital tools that are still in the process of development and refinement, similar to the initial stages of GIS in the 1960s and 1970s, the process of digitizing and drawing "on the fly," as well as evaluating for performance of various scenarios, has improved greatly. In a geodesign process, the design is conceptualized as a collaboration in which there are no owners of ideas and where the computers respond to changes in design as it is being built by various stakeholders. The workflow or the collaborative design process is streamlined, allowing more time for thinking and discussing among participants.

THE STUDY AREA: THE COAST OF GEORGIA AND CHATHAM COUNTY, UNITED STATES

The coast of Georgia represents a socially and ecologically diverse region in the Southeast United States. An aerial photograph of this region provides a distinctive view of its natural and built environments in contrast with other southeastern coastal areas: a vast marshland dominated by Spartina grass (*Spartina alterniflora*) that still covers much of the coastal marsh area of Georgia, with relatively few urbanized areas or concentrations of population and relatively low density (figures 2.4a and 2.4b).

Figures 2.4a and 2.4b. Coastal Georgia aerial views depict marshes and little presence of built areas.

Photos by Alison L. Smith.

The six Georgia counties that make up this region currently contain 378,000 acres of salt marsh, which account for more than one-quarter of the remaining salt marsh along the Atlantic coast. These large tracts of preserved salt marsh are largely the result of Georgia's Coastal Marshlands Protection Act, enacted in 1972 (Seabrook 2012). The Georgia Bight, the long, indented curve of the Atlantic coast that stretches from Cape Hatteras, North Carolina, to Cape Canaveral, Florida, along with other features of the Georgia coast, helps to naturally shelter this region from tropical storms and hurricanes, following the Gulf Stream's warm waters (Davis 2013) (figure 2.5).

Figure 2.5. Map of Georgia's coastal tidal saltwater marshes.

Map by Alison L. Smith; data courtesy of Esri, US Census Bureau.

Chatham County is located in the northern boundary of this coastal region. The largest city is Savannah, with a population of 144,352. It was founded by James Oglethorpe in 1733, on the margins of the Savannah River. This was the first planned settlement in what would become the state of Georgia. Savannah's National Historic Landmark District and eight other historic districts encompass an area of 3.3 square miles, with more than 6,000 architecturally significant buildings. (Chatham County - Savannah Metropolitan Planning Commission 2016). Due to its historic nature, Savannah is a popular tourist destination with a distinct sense of place. However, the original placement of the city and its design could also be considered an example of resilience, based on Oglethorpe's careful surveying of the region and the selection of a bluff (or higher ground) that kept inhabitants safe not only from invaders but also from flooding and associated diseases, such as malaria. As a result, such meticulous planning with the environment enables the historic district to remain above water even under a category 5 storm surge (Alfonso 2014). The basic plan of the city consisted of a ward system (600 by 600 feet), each ward representing a neighborhood by itself, with buildings and streets facing a square. As it happened in many other coastal cities in the world, as the city grew outside its original plat, it expanded into the floodplains and the marshes, representing unsuitable areas for development: the need to expand suppressed the natural boundaries imposed by the natural environment (Alfonso 2014).

Today, Savannah and the rest of Chatham County rely mostly on three main economic activities: the Port of Savannah, the military base (Fort Steward), and tourism. Tybee Island, one of several barrier islands on the coast of Georgia, receives part of the tourist demand, along with other more developed islands such as Jekyll Island and St. Simon. Interstate 95, which runs through the entire Eastern Seaboard (from Miami to Maine), plays an important role in connecting Savannah, the port, and the region with other markets.

Exposure to natural hazards

Like many coastal regions, sea level rise and hurricanes present a prevailing issue. The Georgia coast was fortunate to avoid a major hurricane strike for over 100 years. Three scenarios of vulnerability to natural hazards have been defined, based on storm surge and hurricane categories: scenario 1, the lowest impact, is a tropical storm; scenario 2 is hurricane categories 1 and 2, and scenario 3 is hurricane categories 3–5 (Coastal Regional Commission of Georgia 2015). More than 1.8 million acres (up to 74.5 percent, 1.3 million acres) of the natural vegetation cover could be potentially impacted by a combination of storm surges and inundation from the highest category of hurricane (scenario 3, hurricanes of categories 3–5), and more than 25 percent (490,000 acres) could be impacted under a tropical storm (scenario 1).

Socioeconomic impacts

Chatham County had a population of 265,128 in 2010, with the city of Savannah representing over half (56.08 percent, 144,352 inhabitants) (US Census Bureau 2010). The total population in the six coastal counties in the region is 552,470. Under various tropical storms and hurricane events, estimates indicate between 61 percent and 86 percent of the total population of the six-county region could be impacted (Coastal Regional Commission of Georgia 2015). State of Georgia projections indicate that Chatham County's population will increase by over 53,761 between now and 2050, with the whole six-coastal region increasing by 76,841 by 2050 (Georgia Governor's Office of Planning and Budget 2020) (table 2.1).

Table 2.1. Population of the six coastal counties in the state of Georgia, 2010–2050

County	2010 total population	2020 total population	2050 population projection	Population change 2020–2050
Chatham	265,128	290,390	344,151	53,761
Bryan	30,233	40,478	49,194	8,716
Liberty	63,453	65,088	61,485	−3,603
McIntosh	14,333	14,503	11,784	−2,719
Glynn	79,626	85,971	106,185	20,214
Camden	50,513	56,040	56,512	472
Total region	503,286	552,470	629,311	76,841

Source: US Census (2010), Esri (2020 estimates), Georgia Governor's Office of Planning and Budget (Population Projections 2018–2063).

The economic base of this region is supported by a total of 20,534 businesses and 303,689 employees who produce an annual total in regional retail sales of $9,839,913,109 (table 2.2) (Esri 2020). Around 80 percent, and up to 89 percent, of these total sales could potentially be affected under scenario 3 (hurricanes of categories 3–5), with around 30 percent under the second scenario.

Table 2.2. Economics of the six coastal Georgia counties potentially affected by a storm surge

County	Total retail sales (USD)	Total businesses	Total employees
Chatham	6,039,362,313	12,042	152,524
Bryan	539,026,710	1,057	8,489
Liberty	668,229,288	1,503	53,878
McIntosh	101,478,203	383	2,217
Glynn	1,919,999,183	3,976	52,908
Camden	571,817,412	1,573	33,673
Total for region	9,839,913,109	20,534	303,689

Source: Infogroup Inc. 2020 and Esri Community Analyst.

Infrastructure

Two of the most important ports in the southeastern United States (Savannah and Brunswick) are in this region. The main transportation corridors extending north–south and connecting with the rest of the Eastern Seaboard are Interstate 95 and State Route 17. In the west–east direction, Highway 80 and Interstate 16 are the main connectors (figure 2.6).

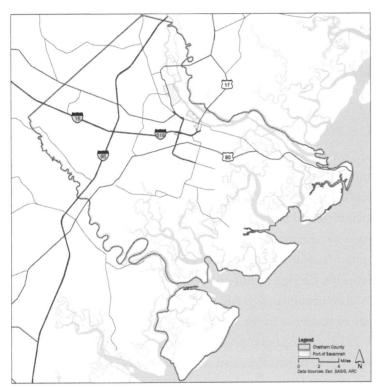

Figure 2.6. Major roads and other infrastructure, including the Port of Savannah, in Chatham County, Georgia.

Map by Alison L. Smith; data courtesy of Esri, Savannah Area GIS (SAGIS), and the Atlanta Regional Commission (ARC).

The primary economic driver increasing development along the Ogeechee River is the Savannah Harbor Expansion Project. The Port of Savannah is the country's fourth-busiest container port and the second-busiest port on the Eastern Seaboard. It recently surpassed Charleston as the primary port of the southeastern corridor. Since 2001, the Port of Savannah has been the fastest-growing container port in the country, which has increased jobs for Chatham County (Ramos 2014). It has been estimated that the port has generated approximately 20,000 distribution jobs for Savannah-area residents (Buntin 2009). Expansion of the Port of Savannah includes dredging the Savannah River to allow larger container ships access to the port.

In terms of other modes of transportation, 4,000 miles of roads could be impacted by the highest level of threat (hurricanes of categories 3–5), representing 66.8 percent of the total mileage of roads in the region. Other infrastructure—such as cell and other communication towers, two major United States ports (Savannah and Brunswick), bridges, shelters, and emergency evacuation routes—needs to be protected to maintain the transportation and communication network required under any of these scenarios.

THE PROBLEM

The Coastal Regional Commission of Georgia (CRC) is the regional planning entity responsible for planning for this 10-county area, reporting to the Georgia Department of Community Affairs (DCA). The region must meet the requirements of DCA for comprehensive planning and to identify critical needs, laying the groundwork for policy and program development during the regional planning process (communication with Lupita McClenning, CRC planning director). On the other hand, the University of Georgia's College of Environment and Design (CED), through its graduate program in environmental planning and design's regional studios, has been addressing issues of resilience in the region as part of a partnership with the CRC initiated in July 2013. Among other projects, through this partnership, one of the regional studio exercises assessed the existing planning tools to address hazard risk and community resilience to be integrated into CRC resilience guidelines and performance standards as a component of the update to the regional plan (Coastal Regional Commission of Georgia 2015).

Other studies, in the form of graduate student thesis projects, were conducted by the University of Georgia CED. A study on how urban form impacts climate and how design could aid the process of adaptation was addressed (Alfonso 2014). This research assessed how climate factors interact with physical landscape, the different climatic responses between the built environment and the natural landscape, what key climate factors have a direct impact on climatic perception and affect comfort, and what design solutions can be examined that could improve the effects of the built environment on climate. The methodological approach considered three scales: regional, city, and site specific.

In March 2015, the effort to assess the resiliency of communities continued with the creation of a resiliency matrix to test the resilience of planning documents for coastal Georgia. This matrix was used as a checklist to evaluate the performance of planning documents for managing the conditions generated by the impact of a natural event and to help in identifying missing portions of documents that need to be completed in the future (Agrawal 2015).

In November 2014, another collaboration was established by CED, with Professor Carl Steinitz and Hrishikesh Ballal, both at The Bartlett Centre for Advanced Spatial Analysis (CASA), University College London. Professor Steinitz is a renowned leader in landscape architecture and urban planning and author of *A Framework for Geodesign* (Esri Press, 2012). The research problem for Georgia's project was in how to implement a digital collaborative framework for decision making based on future scenarios that allowed involvement of many stakeholders at the regional and local levels and that explored issues of spatial representation, evaluation, or assessment; iterative designs and multiple scenarios for the future; and an iterative review process to achieve a single negotiated plan for the future of the region (figures 2.7a and 2.7b).

In exploring these, specific problems related to resilience and climate were to be addressed, including flooding events, sea level rise (SLR), hurricane and storm events, saltwater intrusion, and infrastructure issues.

Scenarios of future SLR at the year 2100 range from a low of 0.3 meters to a high of 2.0 meters associated with collapse of polar ice sheets. Understanding the specific locations at risk for SLR impacts is a high priority in climate change research and adaptation planning. In a recent article, Hauer et al. (2016) explored population at risk in the United States under various SLR scenarios, finding that by 2100, with an SLR of 0.9, 4.2 million people could be at risk of inundation in the United States, with a large increase (three times larger, 13.1 million people) with SLR at 1.8 meters. The authors concluded that protective measures are needed to avoid large displacement of the population at the scale of what occurred with the Great Migration of African-Americans out of the South in the 20th century. Along the coast of Georgia, the total population potentially displaced in Chatham County by a 0.9-meter SLR would be 49,587 (Hauer et al. 2016).

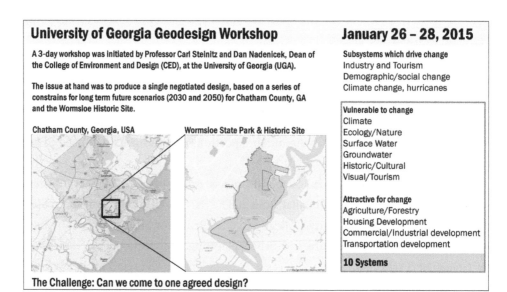

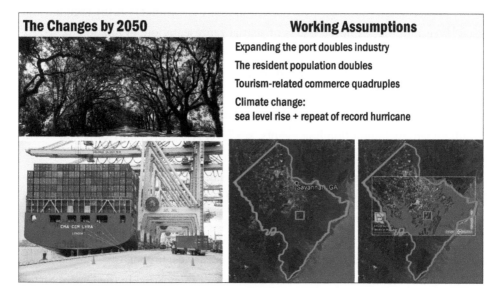

Figures 2.7a and 2.7b. Objectives and working assumptions for the 2015 Coastal Georgia Geodesign Workshop.

Reprinted with permission of University of Georgia and Carl Steinitz. Photos for 7b by Alison Smith (Wormsloe) and USDA (Port of Savannah) (www.flickr.com/photos/usdagov/35078027780).

METHODS

Some of the methods and the data described in this section were previously reported in the proceedings of the Digital Landscape Architecture Conference (Rivero et al. 2015) and a presentation at the 2016 Geodesign Summit in Redlands, California (Smith and Rivero 2016). To set the stage for this section of the chapter, we present our methods at three levels:

A. The overall method to establish the geodesign project and particularly the evaluation matrix and maps used for the implementation of the process

B. Specific methods to generate the 10 evaluation maps: identifying and mapping the 10 systems, as well as suitability of different systems

C. Methods for resilience in general, and particularly for coastal areas, modeling for resilience based on existing models for SLR and storm surge:

- NOAA SLR data, based on current mean higher high-water inundation extent

- Sea, Lake, and Overland Surges from Hurricanes (SLOSH) model

A: General geodesign methodological approach

The methodological framework to follow for a geodesign process is based on Steinitz (2012) and starts with the series of questions presented earlier in the chapter:

1. How should the study area be described?

2. How does the study area function?

3. Is the current study area working well?

4. How might the study area be altered?

5. What difference might the changes cause?

6. How should the study area be changed?

In his geodesign framework, Steinitz (2012) develops answers for these questions through an iterative process. The first three questions are intended to generate an assessment of the region, derived from representation, process, and evaluation models. Next, the second set of three questions is intended to introduce future scenarios of change and to characterize or generate change, impact, and decision models. The study is conducted through three iterations, in which the first iteration provides the research team with an understanding of the context, the second iteration defines and specifies the methods, and the final iteration represents the study itself.

In adopting this framework, the process starts with identifying the problem and the list of 10 issues/systems. In very broad terms, this initial phase includes the following steps (Smith and Rivero 2016):

1. Define the study area and issues.

2. Identify the 10 issues/systems and start the mapping process for the evaluation maps that represent each of the 10 issues/systems.

3. Identify experts to define evaluation map criteria for each issue/system.

4. Create a matrix for the 10 issues/systems to document model criteria, data needs, and so on.

5. Collect GIS data needed for each of the 10 issues/systems.

6. Create evaluation maps for each issue/system in GIS (10 evaluation maps total).

7. Email evaluation maps (as shapefiles) to be uploaded into the Geodesignhub software.

While evaluating the 10 systems, a parallel planning and logistic process starts that involves the identification of potential stakeholders that could represent each of these 10 teams. This phase of the process is intended to cover the wide spectrum of disciplines and areas of any geodesign process, in a sort of "human landscape" mapping. Figure 2.8 shows an overlap between Steinitz's original geodesign diagram (2012) and the list of participants from the 2015 Coastal Georgia Geodesign Workshop.

The workshop planning efforts started in November 2014, with a steering committee composed of two teams:

1. A leading team from CASA, University College London, UK (Carl Steinitz, Hrishikesh Ballal, and Tess Canfield). The lead team's role was to guide the process and conduct the geodesign workshop, while also implementing and providing support and training for the use of the Geodesignhub software (https://www.geodesignhub.com).

2. A local team from the University of Georgia (team leaders: Rosanna Rivero and Alison Smith). The local team was responsible for workshop planning and coordination, identification of participants and teams, definition of

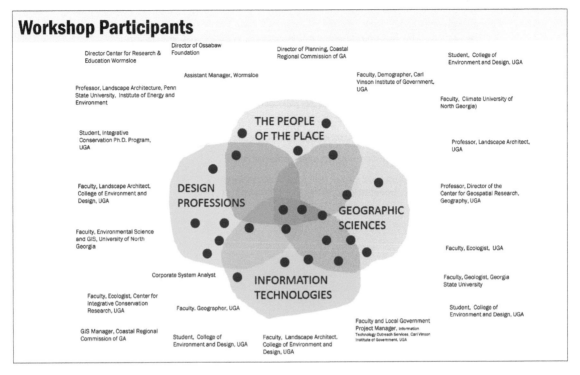

Figure 2.8. The human geodesign landscape: Participants in the Coastal Georgia Geodesign Workshop and their location in Steinitz's geodesign people diagram.

Printed with permission from Rivero et al. 2015.

study area and issues, GIS data preparation, and creation of all evaluation maps. This team had the knowledge of the area, the stakeholders to participate in the workshop, and also the facilities and support for organization.

Bringing together both process experts and local experts produces a stronger team to develop the trust and participation of local stakeholders. This formula has been used by Steinitz and Ballal through Geodesignhub in other case studies, in some cases complementing remote experts on particular topics, such as ecological modeling (Rivero et al. 2016).

The Geodesignhub online application (https://www.geodesignhub.com) is intended to create a digital workflow that participants can learn to use during the workshop, after a short tutorial, and during the various phases of the geodesign workshop process (Ballal et al. 2016). Figure 2.9 provides a synthesis of the digital workflow.

Software: Digital Workflow for Dynamic Geodesign Synthesis   Hrishi Ballal

- a digital web based workflow to support the rapid creation of conceptual designs to address large and complex geodesign problems

- designed to foster collaboration between professionals and their clients, and among teams of professionals, during the early stages of design

- simple user interface which easily incorporates existing and diverse data structures for both its inputs and outputs

- enables users to collaborate in person and/or over the internet in real time to produce designs and assess them.

- the tool is publically available and free to use for all. It has extensive self-help documentation, with video and text tutorials. These are available by signing up for an account at geodesignhub.com.

○ GEODESIGN HUB

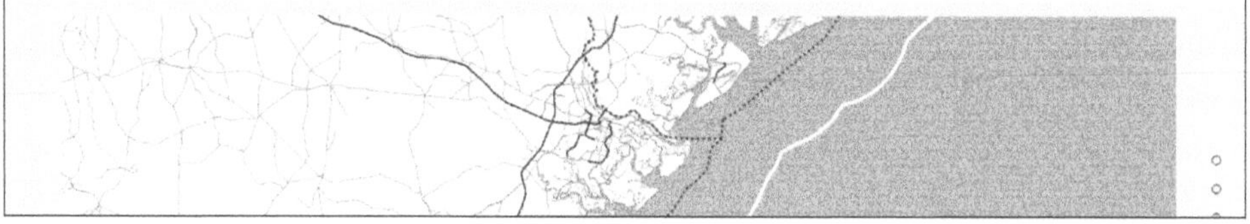

Figure 2.9. Synthesis of the digital workflow of the Geodesignhub web application.

Reprinted with permission of Hrishikesh Ballal.

B: Specific methods to generate the 10 evaluation maps

After defining the study area and identifying the 10 systems and experts to define the evaluation map criteria for each system, a shared network folder was created for the workshop, allowing geographically dispersed teams and system experts to collaborate. A matrix was created in a spreadsheet for the evaluation maps to identify the criteria to be considered, a specific person responsible for identifying the criteria, data needs, sources, and other supporting information (Smith and Rivero 2016).

As Steinitz (2012) explains, evaluation maps result from a need and a judgement, based on input from stakeholders and decision makers, to assess the comparative impacts between the present state and the future or changed state that may develop following any proposed design. Each evaluation map is defined as either a vulnerability map or an attractiveness map and consists of three levels represented by colors (high is red, medium is yellow, and low is green). "Evaluation criteria are typically expressed in relation to their positive characteristics, such as their attractiveness for a particular purpose, or for their negative characteristics, including their contribution to the vulnerability of a particular resource, location, or action" (Steinitz 2012).

In our 2015 workshop, the criteria considered for each map were drawn from expert knowledge and available GIS data. Data layer needs were added to the matrix in consultation with the planning team and then collected to map these criteria. This was a complex task that required several iterations and consultations with experts in each system. For the Georgia project, the primary source for GIS data was the Savannah Area Geographic Information System (SAGIS) website: www.SAGIS.org. Other data sources were also used (Rivero et al. 2015).

The 10 systems identified for the study area were the result of many discussions within the planning team, along with consultation with groups of experts. These 10 systems include climate, nature/ecology, surface water, groundwater, historic/cultural resources, visual resources, agriculture/forestry, housing, commercial/industrial, and transportation/evacuation routes (figure 2.10).

The GIS mapping methodology was streamlined using ArcGIS® software to facilitate easy integration into the Geodesign software. What follows is a workflow for the creation of the evaluation maps, using ArcGIS software, with a more refined explanation (in the next subsection) of how maps for climate were created. In this subsection, the commercial/industrial evaluation map is used as an example to illustrate the process for the 10 evaluation maps (Smith and Rivero 2016):

As mentioned earlier, a matrix was created and shared for each of the 10 systems so each group of experts could identify and define the criteria to assess levels of attractiveness or vulnerability, data needs, data sources, and other parameters. The first step in the creation of the evaluation maps was referencing the criteria and GIS data layer needs identified in the matrix. An example of one such matrix for the commercial/industrial evaluation map is shown in figure 2.11.

For the ArcGIS Initial Compilation map, all data layers were compiled and gathered into group layers in ArcMap™ by color/assessment level (figure 2.12).

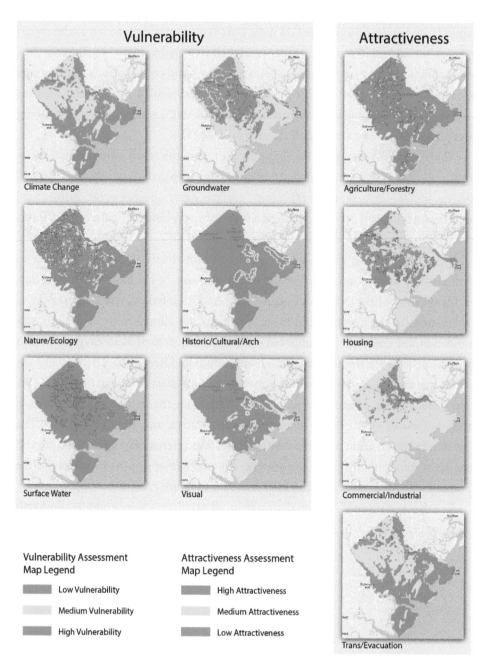

Figure 2.10. Ten evaluation maps for 2015 Coastal Georgia Geodesign Workshop, Chatham County.

Figure by Alison L. Smith.

CRITERIA		
Low Attraction (red)	**Med Attraction (yellow)**	**High Attraction (green)**
Existing commercial/industrial areas; swamp/marsh areas; conservation lands	All other areas/ areas that aren't currently legally available but could be built as commercial/industrial	Areas most attractive for commercial/industrial but not currently a comm/ind use. - areas identified as comm/ind in the FLU plan but not currently a comm/ind use. ; - areas zoned commercial/industrial but not currently a commercial/industrial use
GIS Data layers needed:	**GIS Data layers needed:**	**GIS Data layers needed:**
SAGIS_Parcel2011_LUCodeDes_JOIN (ExistingComm/Indsutrial); HYDRO_hydrology_poly (swamp/marsh); ENVIRO_Conservation_Land	BND_Chatham	BND_Future_Landuse (only Comm/Industrial); BND_Zoning (only Comm/Industrial)

Figure 2.11. Example of commercial/industrial evaluation matrix.

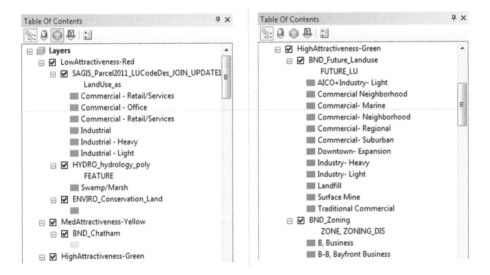

Figure 2.12. Compilation of GIS layers by evaluation color code.

Figure by Alison L. Smith.

For Color/Assessment Level extraction, for each color/assessment level, the following steps were applied: (Commercial/Industrial color/level example: red— low attractiveness)

1. Select by attributes and export only needed features to a new data layer. For Commercial/Industrial, this included selecting all commercial and industrial uses from the existing land use data layer, selecting swamp/ marsh from the hydrology data layer, and exporting each to a new data layer.

2. Dissolve any features if needed. Use the dissolve tool for the following three data layers: Commercial/Industrial Land Uses, Swamp/Marsh, and Conservation Lands.

3. Create a shapefile for each level/color with attributes for Type and Color.

4. Copy dissolved features from each layer that is needed into this new shape- file. Start an edit session and paste all features from the three data layers listed in step 2.

5. Use the Simplify Polygon and Aggregate Polygons tools to generalize features.

6. Combine the three color/evaluation levels into one shapefile.

7. Use the county boundary polygon layer and multiple iterations of the Erase and Union tools to combine the three colors/levels and ensure there are no overlaps.

   The order of this process is important. For example, the existing commercial/industrial areas (red–low attractiveness) will most likely be identified in the Future Land Use plan data layer (green–high attractiveness) as areas for commercial/industrial. You must ensure that the existing commercial/industrial areas are removed from the Future Land Use plan data layer.

8. Use the Simplify Polygon tool to further generalize the combined features and to resolve any topological errors.

9. For the final reduced evaluation map, each evaluation map was sent to the Geodesignhub software team (Ballal) to be uploaded into the Geodesignhub software.

Figure 2.13 shows a summary of the 10 evaluation maps and corresponding matrices following this methodology.

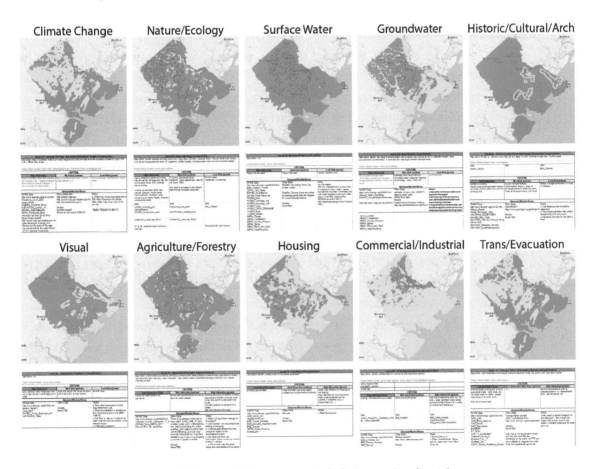

Figure 2.13. A summary view of the 10 evaluation maps with their associated matrices. A slightly different method was used for the climate assessment map, as described in the following subsection.

Figure by Alison L. Smith.

C: Evaluation maps for climate based on existing models for sea level rise and storm surge

Two sources of data were used to create the evaluation maps for climate: the sea level rise projections, generated by NOAA, and the storm surge derived from predicted hurricane models developed by the United States National Weather Service (NWS).

NOAA SLR: We identified the areas at risk of SLR by using NOAA's 0-meter through 1.8-meter (6 feet) SLR datasets (NOAA 2010). These datasets simulate expected changes in the mean higher high water (MHHW) mark on areas that are hydrologically connected to coastal areas, without considering additional land loss caused by other natural factors such as erosion (Hauer et al. 2016). The mapping process used to produce this data can be described as a single-value surface model, also called the modified bathtub model. This approach addresses only two variables: the inundation level and the ground elevation. More complex hydrologic and geomorphic models, which include additional variables, can also be used to depict inundation. For example, detailed models might attempt to account for both local and regional tidal variability and hydrologic connectivity. However, this level of detail is not necessary for regional models in which the objective is to get an overall estimate of the potential impacts.

The process uses two source datasets to derive the final inundation rasters and polygons and accompanying low-lying polygons: the digital elevation model (DEM) of the area and a tidal surface model that represents spatial tidal variability using a limited number of closest gauge stations when interpolating the tidal values across the area. The tidal model is created using NOAA's National Geodetic Survey VDatum datum transformation software (http://vdatum.noaa.gov) in conjunction with spatial interpolation and extrapolation methods and represents the MHHW tidal datum in orthometric values (North American Vertical Datum of 1988). The second source of data, and also of uncertainty, in the model is the DEM, generated from lidar data, with vertical accuracies that can vary from 5 centimeters (root mean square error, or RMSE) to more than 30 centimeters (RMSE) and may vary within any one collection area (NOAA 2010).

Figure 2.14 shows the levels of confidence of SLR and coastal flooding projections for Chatham County, with blue areas denoting locations mapped as a high degree of confidence (80 percent) that they will be flooded and unshaded areas denoting a high confidence that these areas will be dry, given the chosen water level. Areas in orange are those with a high degree of uncertainty representing locations that may be mapped correctly (either as inundated or dry) fewer than 8 out of 10 times (https://coast.noaa.gov/slr).

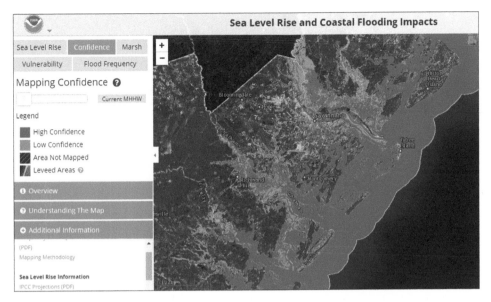

Figure 2.14. Sea level rise and coastal flooding for Chatham County, Georgia.

Retrieved from NOAA online viewer at https://coast.noaa.gov/slr.

Hurricane SLOSH model

The Sea, Lake, and Overland Surges from Hurricanes (SLOSH) model is a computerized numerical model developed by NWS to estimate storm surge heights resulting from historical, hypothetical, or predicted hurricanes by considering the atmospheric pressure, size, forward speed, and track data. These parameters are used to create a model of the wind field, which drives the storm surge. These models are disseminated by NWS and the National Hurricane Center (NHC) (www.nhc.noaa.gov /surge/slosh.php) for various purposes (NHC 2016).

The SLOSH model consists of a set of physics equations, which are applied to a specific locale's shoreline, incorporating the unique bay and river configurations, water depths, bridges, roads, levees, and other physical features. NWS uses a composite approach, which is a mix of deterministic and probabilistic approaches, to predict surge by running SLOSH several thousand times with hypothetical hurricanes under different storm conditions. The products generated from this approach are the Maximum Envelopes of Water (MEOWs) and the Maximum of MEOWs (MOMs), which are regarded by NHC as the best approach for determining storm surge vulnerability for an area since they take into account forecast uncertainty. The MEOWs and MOMs play an integral role in emergency management, as they form the basis for the development of the nation's evacuation zones (NHC 2016).

Processing of climate data (combining SLR and storm surge) for 2015 Coastal Georgia Geodesign Workshop

The two sources of data described earlier—NOAA sea level rise and storm surge—were used to address the climate evaluation maps for the 2015 Coastal Georgia Geodesign Workshop. The general processing for each of these datasets was as follows:

1. Download the datasets from the NOAA SLR and NHC websites.

2. Clip both datasets to the Study Areas boundary (Chatham County).

3. Use the matrix shown in table 2.3 to intersect both datasets and identify areas to be considered of high vulnerability (red), medium vulnerability (yellow), or low vulnerability (green).

4. Erase all areas in both datasets from the study area to obtain remaining areas in the study area that would represent yellow areas (areas not affected by either SLR or storm surge).

5. Merge these erased areas back to the clipped file to generate one dataset.

6. Merge each category's green (remaining areas) with the red and yellow areas.

7. Use the Intersect function to intersect both SLR and storm surge.

8. Use the Add a new numeric attribute field in each table called CLASS and add a new value (1 to 3) in the CLASS field (table 2.3). This would be the equivalent to using the Reclass function in a raster-based spatial analysis operation.

Table 2.3. New numeric values assigned to each SLR and storm surge category

SLR	Class	Storm
1 ft	1	Tropical storm and category 1 hurricane
2–3 ft	2	Category 2–3 hurricane
9 ft	3	Category 4–5 hurricane

9. Add another numeric attribute field to perform the operations shown in table 2.4 from the reclassified values from the previous step for each attribute field.

Table 2.4. Matrix used to intersect sea level rise and storm surge categories with color values

Storm surge	Sea level rise		
	1 ft (value = 1)	2–3 ft (value = 2)	Other (value = 3)
Cat 1 (value = 1)	1 + 1 = 2 (red)	1 + 2 = 3 (red)	1 + 3 = 4 (yellow)
Cat 2-3 (value = 2)	2 + 1 = 3 (red)	2 + 2 = 4 (yellow)	2 + 3 = 5 (yellow)
Cat 4-5 (value = 3)	3 + 1 = 4 (yellow)	3 + 2 = 5 (yellow)	3 + 3 = 6 (green)

10. Reduce file: Use the Final ADD and RECLASS fields to convert to raster with 10 meters (32.80 feet).

11. To upload the map to the Geodesignhub online application, the final shapefile had to be reduced in size and number of features. To perform this operation, we aggregated and generalized the surface in the final stages of the process. A tool to automate this operation was developed using ModelBuilder™, as shown in figure 2.15. This operation involved the following steps:

a. Exported the final file to a lower-resolution raster (32.8 ft × 32.8 ft, resampled).

b. Using the Spatial Analyst tools, we applied a generalization tool (Boundary Clean) and sorted with descend priority.

c. The raster was converted back to a polygon using the Raster to Polygon (Conversion) tool.

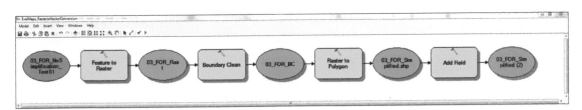

Figure 2.15. An example from a tool developed in ModelBuilder to generalize and reduce the file size to upload in the Geodesignhub online application.

BEST PRACTICES AND LESSONS LEARNED FROM THE PROCESS

A synthesis of the results from the design process during the three-day workshop is presented in figure 2.16. Preliminary results from the process of creation of the 10 evaluation maps were already presented in the methods section. This figure also presents the results from five change teams that were generated later in the process to create designs under each group of stakeholders' perspectives. The ultimate goal was to produce a single negotiated proposal. Figure 2.17 shows a synthesis of the online application after data was uploaded and some of the metrics were applied, and figure 2.18 shows the Geodesignhub interface.

The geodesign workshop is the result of an interdisciplinary process involving participants with varied levels of knowledge about either the study area or the topics of study. It also represents a cross-pollination between academia and practice. A contemporary framework to address resilience requires this type of collaborative, interdisciplinary approach, given the complexity and high level of interdependence between issues. Given the cross-disciplinary nature associated with planning for resilient communities, these topics can no longer be addressed in disciplinary silos. To better understand and integrate the relationships between issues such as climate, water, and conservation and their effects on the built environment in future iterations of such workshops, we have begun to move forward to incorporate issues of green infrastructure and ecological connectivity.

It has already been recognized by other authors in the geodesign arena that "the specific ingredients of each project will depend on the issues, participants, available data, information, knowledge, culture, values, geographic context, and available technology" (McElvaney 2012). There is a steep learning curve during the initial phases of such workshops due to a combination of factors: the complexity of the online tool itself and the added complexity of social/human interactions among the stakeholders. The workshop provided an environment where participants entered the process without a common language, knowledge, or area/field of interest but, through the geodesign process, used a combination of local and expert knowledge together with the capabilities of GIS to provide a visual and iterative view of potential options to initiate an expedited method to work within and between the various teams (Rivero et al. 2015).

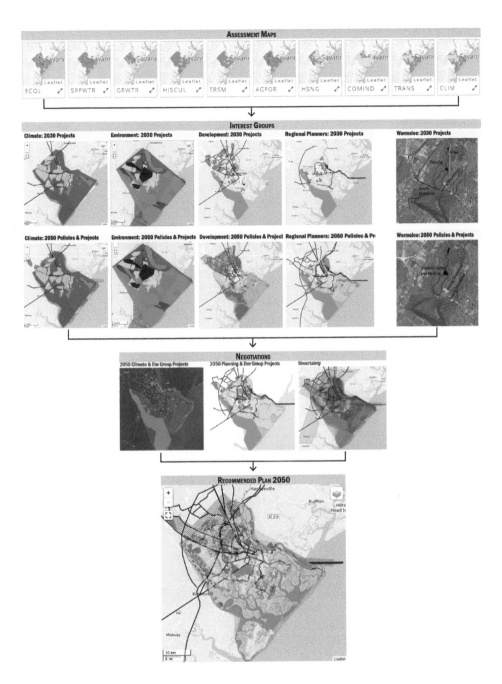

Figure 2.16. Results from the geodesign process for Chatham County, Georgia.

Photos reprinted with authorization from Carl Steinitz and Hrishikesh Ballal.

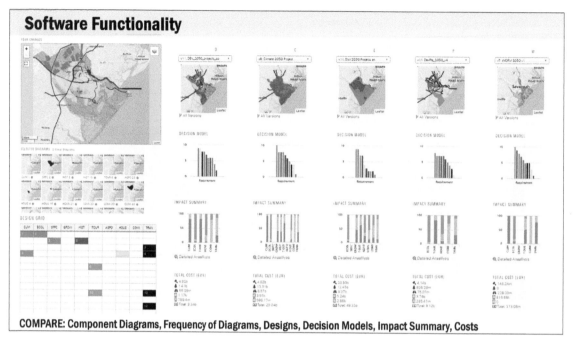

Figure 2.17. Synthesis of the Geodesignhub software dynamic functionality.

Photos reprinted with authorization from Carl Steinitz and Hrishikesh Ballal.

Figure 2.18. The Geodesignhub interface, along with photos of conductors and participants of the workshop.

Photos reprinted with authorization from Carl Steinitz and Hrishikesh Ballal.

Specific aspects of these study areas, such as limitations on areas available to address population growth, are common in coastal areas around the world that are experiencing similarly rapid development. One of the workshop participants stated:

> Chatham County can accommodate residential growth through increasing density in and near existing residential areas. However, since much of the warehousing and industrial development associated with the port will be single story, it will be impossible to accommodate the required area within the boundaries of Chatham County. As well as protecting its environment and consolidating increased development in its northwestern area, it will be necessary to expand development outside Chatham County. This will mean growth in adjoining Georgia counties and the need for increasing cooperation and collaboration in planning with South Carolina, areas which will also benefit from the growth of the Savannah region. (Brian Orland, extract from the workshop)

Another participant, Lupita McClenning, director of planning and government services at the Coastal Regional Commission, recognizes the potential application of this process in the future in areas beyond this region, noting that by 2050 we will be responding to a changing world and a changing population (Rivero et al. 2015). According to McClenning, long-term coordination is required between land-use controls and public and private investments on the local, regional, and cross-state scales to be effective and efficient. McClenning was instrumental in pursuing funds to organize a second workshop in April 2016, this time involving the 10-county coastal region of Georgia, and with the challenge of generating a multiscale and multijurisdictional implementation of the process and the tools (Rivero et al. 2016).

Finally, Hauer et al. (2016) provide estimates of annual global costs for coastal adaptation strategies for flood protection infrastructure on the order of $421 billion (USD) per year to address a predicted sea level rise of 2.0 meters by 2100 (2014 values). These numbers do not consider the expansions in population and infrastructure that are likely to take place before one of these scenarios of inundation occurs. Other strategies, such as managed retreat solutions, could also prove troublesome if population projections and other future changes are not considered in future planning, particularly for larger cities or areas with expected growing populations.

According to the authors, not only could the costs of relocating a community be greatly underestimated if that population is growing, but the challenge of finding suitable areas for relocation could be problematic as well. With current estimates as high as $1 million (USD) per resident in some small Alaskan villages, each decade both increases that population's exposure to SLR and increases its vulnerability to the economic costs of inaction. Potential growth management strategies in high-risk areas experiencing rapid population growth could also prove more effective than relocation (Hauer et al. 2016). With these ideas in mind, techniques, methods, and tools—such as those presented in this chapter to help address future changes in a collaborative and highly interdisciplinary environment—will be necessary to provide better estimates that can be used by decision makers to plan for resilience in coastal communities facing these complex challenges.

CONCLUSION

The geodesign approach can be applied to resilience and other types of projects, as long as there is (1) a clear definition of the issues at hand and potential future changes, (2) one or more leading groups with not only enough understanding of these issues for a particular region but also with the capacity to coordinate and bring to the table a diverse and representative group of stakeholders, and (3) a minimum access to knowledge of the resources of a region—enough to map them, ideally with the use of technology (GIS/geospatial tools). Mapping and modeling resources with these technologies is ideal but not a strict requirement, as some of these tools (such as the Geodesignhub software used for this particular project) allow digitizing on the screen, offering the possibility to collect information from people with the knowledge of the area. Finally, a clear understanding of the current situation and the criteria to establish optimal capacity, feasibility, and suitability of the land by consultation with experts from the place are some of the key requirements for a successful geodesign project.

It is important to mention that two primary and essential functionalities of the Geodesignhub software include the ability to draw and integrate diagrams into a design scenario and the ability to evaluate the impact of a design scenario in real time. Real-time impact evaluation allows participants to generate multiple design scenarios throughout the workshop, refining previous scenarios based on impact evaluation and negotiations with other county and regional teams.

Results of this project have already been used to improve planning efforts by Carl Steinitz and Hrishikesh Ballal in other regions, including Washington State; Italy; and Belo Horizonte, Brazil. Additionally, as part of CRC and the University of Georgia's ongoing collaboration agreement, and also CRC future monitoring plans for the region along with education and outreach, there are plans to continue using these tools to train more planners and other professionals. It is our hope that this chapter will also encourage many other regions to implement a similar approach.

ACKNOWLEDGMENTS

The authors thank Carl Steinitz, Hrishikesh Ballal, and Tess Canfield for their guidance, the framework and software development, and the assistance with and implementation of the Geodesignhub tool in this project; the other UGA workshop planning team members (Sarah Ross, Jon Calabria, Alfie Vick, Marguerite Madden, and Tommy Jordan); and the workshop participants. Funding for this project was from the College of Environment and Design at the University of Georgia, the Wormsloe Foundation, and the Rado Family Foundation.

REFERENCES

Agrawal, S., 2015. "Comprehensive Definition of Resilience in Urban Planning." Unpublished master's thesis, University of Georgia, Athens, GA.

Alfonso, M. B., 2014. "Planning with Climate: Urban Design as a Tool for Adaptation." Unpublished master's thesis, University of Georgia, Athens, GA.

Biggert-Waters Flood Insurance Reform Act of 2012. Public Law 112–141; 126 Stat. 957 (codified as amended at 42 U.S.C. §§ 4001–4131).

Binita, K. C., J. M. Shepherd, and C. J. Gaither. 2015. "Climate Change Vulnerability Assessment in Georgia." *Applied Geography*, 62: 62–74.

Blake, Eric S., Christopher W. Landsea, and Ethan J. Gibney. 2011. *The Deadliest, Costliest, and Most Intense United States Tropical Cyclones from 1851 to 2010.* NOAA Technical Memorandum, National Weather Service (NWS) NHC-6. Miami, FL.

Buntin, J. 2009. "Charleston, SC, and Other Cities Protect Their Urban Waterfronts." In *Cities and Water: A Handbook for Planning*, edited by Roger L. Kemp, 57–62. Jefferson, NC: McFarland.

Byrne, Kevin. 2015. " 'King Tide' Spurs Coastal Flooding from Charleston, SC, to Tybee Island, GA." AccuWeather.com, last modified October 30, 2015. www.accuweather.com/en/weather-news/photos-king-tide-spurs-coastal-flooding-from-charleston-sc-to-tybee-island-ga/379846.

Chatham County - Savannah Metropolitan Planning Commission (MPC). 2006. "Tricentennial Plan Chatham County – Savannah Comprehensive Plan Community Assessment Report." Accessed November 15, 2015. www.thempc.org/docs/lit/compplan/2012/dec/communityassessment.pdf.

Chatham County - Savannah Metropolitan Planning Commission. 2016. "Chatham County–Savannah Comprehensive Plan: 2016 Update."

Claritas Nielsen. 2013. "Business Facts 2013." Report generated for six Georgia counties.

Coastal Regional Commission of Georgia. 2015. "Regional Assessment of Coastal Georgia." Darien, GA.

Cutter, S. L., B. J. Boruff, and W. L. Shirley. 2003. "Social Vulnerability to Environmental Hazards," *Social Science Quarterly* 84 (2): 242–61.

Cutter, S. L., and C. Finch. 2008. "Temporal and spatial changes in social vulnerability to natural hazards." *Proceedings of the National Academy of Sciences*, 105 (7): 2301–6.

Dangermond, J. 2010. "GeoDesign and GIS—Designing Our Futures." *Peer Reviewed Proceedings of Digital Landscape Architecture*, edited by Buhmann, Pietsch, and Kretzler. Anhalt University of Applied Sciences. Berlin/Offenbach: Wichmann. 502–14.

Davis, J. 2013. *Island Time: An Illustrated History of St. Simon Island, Georgia.* Athens, GA: University of Georgia Press.

Eraydin A., Taşan-Kok T. 2013. "The Evaluation of Findings and Future of Resilience Thinking in Planning." In Eraydin A., Taşan-Kok T. (eds) "Resilience Thinking in Urban Planning." GeoJournal Library, vol 106. Springer, Dordrecht, 229–39. https://doi.org/10.1007/978-94-007-5476-8_14.

Ervin, S. M. 2016. "Technology in geodesign." *Landscape and Urban Planning*, 156: 12–16.

Esri. 2020. "Community Analyst: Population Estimates and Business Statistics" (data file).

Evans, Jason, Rob McDowell, Chuck Hopkinson, Jill Gambill, David Bryant, Kelly Spratt, and Wick Prichard. 2013. "Tybee Island Sea Level Rise Adaptation Plan." Athens, GA: Georgia Sea Grant, Carl Vinson Institute of Government, City of Tybee Island.

Flaxman, M. 2010. "Fundamentals of geodesign." *Peer Reviewed Proceedings of Digital Landscape Architecture*, edited by Buhmann, Pietsch, and Kretzler. Anhalt University of Applied Sciences, Berlin/Offenbach: Wichmann.

Galloway, Jim. 2015. "On the Georgia coast, climate change has become more than theoretical." *Atlanta Journal Constitution*, October 31, 2015. www.ajc.com/news/news/state-regional-govt-politics/on-the-georgia-coast-climate-change-has-become-mor/npCpB.

Geodesignhub. n.d. Accessed October 13, 2019. https://www.geodesignhub.com/#collab.

Governor's Office of Planning and Budget, Performance Management Office. 2010. Georgia 2030 – Population Projections. Accessed November 15, 2015. www.georgialibraries.org /lib/construction/georgia_population_projections_march_2010.pdf.

Governor's Office of Planning and Budget, Performance Management Office. 2012. Georgia Residential Population Projections: 2010 – 2030 (data file).

Gannon, Michael. 1993. *Florida: A Short History.* University Press of Florida.

Grunwald, Michael. 2007. *The Swamp: The Everglades, Florida, and the Politics of Paradise.* Simon and Schuster.

Hauer, Mathew E., Jason M. Evans, and Deepak R. Mishra. 2016. "Millions projected to be at risk from sea-level rise in the continental United States." *Nature Climate Change* advance online publication. https://www.nature.com/articles/nclimate2961.

Holling, C. S. 1973. "Resilience and Stability of Ecological Systems." *Annual Review of Ecology and Systematics* 4: 1 – 23.

Kleinberg, Eliot. 2004. *Black Cloud: The Deadly Hurricane of 1928.* Carroll & Graf Publishers.

Landers, Mary. 2014. "Tidal flooding forecast as new normal for Savannah, Tybee Island." *Savannah Morning News*, October 18, 2014. http://savannahnow.com/news/2014-10-18 /tidal-flooding-forecast-new-normal-savannah-tybee-island.

Louisiana Recovery Authority (LRA). 2007. "Louisiana Speaks Regional Plan." In *Vision and Strategies for Recovery and Growth in South Louisiana*, edited by Louisiana Recovery Authority.

McElvaney, Shannon. 2012. *Geodesign: Case Studies in Regional and Urban Planning.* Redlands, CA: Esri Press.

McHarg, Ian L. 1995. *Design with Nature.* New York: John Wiley.

Mendes, J. M., and A. Tavares. 2009. "Building Resilience to Natural Hazards: Practices and Policies on Governance and Mitigation in the Central Region of Portugal." *Safety, Reliability and Risk Analysis*, 2: 1577 – 84.

Müller, Bernhard. 2011. "Urban and regional resilience – A new catchword or a consistent concept for research and practice?" In *German Annual of Spatial Research and Policy 2010*, 1 – 13. Springer.

National Hurricane Center. 2016. Sea, Lake, and Overland Surges from Hurricanes (SLOSH) National Oceanic Atmospheric Administration (NOAA). Accessed October 15, 2015. www.nhc.noaa.gov/surge/slosh.php.

National Oceanic and Atmospheric Administration (NOAA) Coastal Services Center. 2010. "Mapping Inundation Uncertainty." South Carolina. https://coast.noaa.gov/data/digitalcoast/pdf/mapping-inundation-uncertainty.pdf.

NOAA National Ocean Service. 2013. *National Coastal Population Report: Population Trends from 1970 to 2020.* NOAA State of the Coast Report Series. US Department of Commerce, Washington, DC.

NOAA. 2016. "Tropical Cyclone Climatology." www.nhc.noaa.gov/climo/#returns.

NOAA. 2020. "Costliest U.S. Tropical Cyclones."
https://www.ncdc.noaa.gov/billions/dcmi.pdf.

NWS. 2015. "The October 2015 Historic Rainfall and Flood Event in Southeast South Carolina." Weather Forecast Office, last modified October 28, 2015. www.weather.gov/chs/HistoricFlooding-Oct2015.

Orland, B. 2015. "The Path to Geodesign: The Family Car of Digital Landscape Architecture?" *Digital Landscape Architecture*, 32–41.

Ramos, Stephen J. 2014. "Planning for competitive port expansion on the U.S. Eastern Seaboard: The case of the Savannah Harbor Expansion Project." *Journal of Transport Geography* 36: 32–41.

Rivero, Rosanna, Alison Smith, Hrishikesh Ballal, and Carl Steinitz. 2015. "Promoting Collaborative Geodesign in a Multidisciplinary and Multiscale Environment: Coastal Georgia 2050, USA." *Journal of Digital Landscape Architecture* 1 (1): 42–58.

Rivero, Rosanna, Alison Smith, Brian Orland, Jon Calabria, Hrishikesh Ballal, Carl Steinitz, Ryan Perkl, Lupita McClenning, and Hunter Key. 2016. "Multiscale and Multijurisdictional Geodesign: The Coastal Region of Georgia, USA." Geodesign Summit Europe, Delft, Netherlands.

Seabrook, Charles. 2012. *The World of the Salt Marsh: Appreciating and Protecting the Tidal Marshes of the Southeastern Atlantic Coast.* Athens, GA: The University of Georgia Press.

Smith, Alison, and Rosanna Rivero. 2016. "Designing the Future of Coastal Georgia with Geodesign Technologies." Esri Geodesign Summit, Redlands, CA.

Steinitz, Carl. 2012. *A Framework for Geodesign: Changing Geography by Design.* Redlands, CA: Esri Press.

Steinitz, Carl. 2016. Personal communication.

United States Congress, House. 2014. Homeowner Flood Insurance Affordability Act of 2014, H.R. 3370, 113th Congress (2013–2014). Accessed March 21, 2014. www.congress.gov/bill/113th-congress/house-bill/3370.

United States Environmental Protection Agency (EPA). 2016. "Smart Growth Strategies for Disaster Resilience and Recovery." Accessed May 30, 2016. www.epa.gov/smartgrowth/smart-growth-strategies-disaster-resilience-and-recovery.

Walker, B., S. Carpenter, J. Anderies, N. Abel, G. Cumming, M. Janssen, L. Lebel, J. Norberg, G. D. Peterson, and R. Pritchard. 2002. "Resilience Management in Social-Ecological Systems: A Working Hypothesis for a Participatory Approach." *Conservation Ecology*, 6(1).

→ CHAPTER 3

Building resilient regions: Spatial analysis as a tool for ecosystem-based climate adaptation

LAUREL HUNT, MICHELE ROMOLINI, AND ERIC STRAUSS

INTRODUCTION

In this chapter, we present the work of Loyola Marymount University's Center for Urban Resilience (CURes), an interdisciplinary research institute in Los Angeles. Researchers, educators, and other professionals at CURes use spatial analysis to better understand and explain patterns in urban areas that impact their overall resilience. In this way, GIS can be a tool to promote and support resilient cities.

Ecologists, social scientists, urban planners, and K–12 educators make up the staff of the Center for Urban Resilience. We are tasked with the CURes mission to use urban ecology to empower communities. Within this broad mission, we conduct projects that are inspired by each of our professional lenses but that, by their nature, are interdisciplinary and integrated. Maps and spatial analyses are one way for us to bring our projects together and to best communicate our research to the local, regional, and international communities we serve.

In this chapter, we describe how we use spatial analysis to support our Mediterranean City Climate Change Consortium (MC-4). We also discuss how this fits into our mission and approach to urban resilience at a variety of scales. MC-4 oversees projects that employ spatial analysis tools to examine ecosystem resilience (figure 3.1). Specifically, the consortium concentrates its efforts on the five similar Mediterranean-climate regions of the world:

- California
- Central Chile
- Western Cape of South Africa
- South Australia and South West Australia
- Region bordering the Mediterranean Sea (including parts of Europe, North Africa, and the Middle East)

Figure 3.1. Mediterranean City Climate Change Consortium (MC-4) regions use spatial analysis tools to examine ecosystem resilience.

Map courtesy of Lisa Pompelli and Dr. Philip Rundel, University of California, Los Angeles.

The Mediterranean climate, moderated by cool offshore ocean currents, is characterized by hot, dry summers and mild, rainy winters. The regions sharing these climate characteristics are located roughly between 30 and 45 degrees latitude on both sides of the equator and represent less than 3 percent of the world's land area. Despite covering a relatively small percentage of the earth's surface, these regions contain globally important resources and infrastructure. For example, they encompass biodiversity hotspots with 20 percent of the world's total plant biodiversity, including over 26,000 endemic species. Additionally, these regions are made up of subnational jurisdictions (cities, states, and provinces) that are strong economic centers with growing populations and tourism as a key driver. For instance, if California were its own nation, it would be the sixth-largest economy in the world, placing it ahead of France and India (Perry 2016).

The similarities in climate mean that the Mediterranean regions also face common challenges as the climate changes. For example, they have limited water availability, are confronting new fire management concerns, and are trying to develop sustainable industry and resources to use while losing natural habitats to agriculture and development. Although the five regions have differing development priorities, their similar climate challenges mean that they present the preeminent opportunity for collaboration and facilitating regionally relevant responses.

MC-4 members are responding to climate change in these five regions through a diverse network of over 100 professionals and academics. By nature, the consortium's work is interdisciplinary; it promotes building resilient communities through collaboration across professional sectors and academic disciplines. Members work in fields as varied as water supply, alternative energy development, green infrastructure design, and human health risk management. Additionally, academic researchers in the network come from a range of departments, such as epidemiology, biology, civil and environmental engineering, urban and regional planning, law, economics, and geography.

Analyzing climate change at the ecosystem scale and connecting it to other similar ecosystems is a novel approach to adaptation. Traditionally, climate change adaptation has happened within governance boundaries using mechanisms such as citywide and statewide plans and policies. Instead of confining its work to city and state boundaries, MC-4's reach extends across these boundaries to incorporate entire ecosystems. Consequently, instead of planning for climate change impacts in the city of Los Angeles, MC-4 plans for the city of Los Angeles in the context of the rest of the Mediterranean ecosystem extending from Northern California in the United States through Baja California, Mexico. Using this lens for planning allows network members to ask questions such as, How do we adapt the entire Santa Ana River Watershed to climate change?

Fortunately, this approach has proved to be advantageous for professionals and researchers tasked with working along traditional governance boundaries as well as those who take an ecosystem-wide approach. During the summer of 2012, MC-4 convened over 230 experts from 15 countries at the inaugural The Mediterranean

City: A Conference on Climate Adaptation in Los Angeles. Participants came from Australia, Mexico, France, Spain, Italy, South Africa, Kuwait, Morocco, Saudi Arabia, Tunisia, Greece, Fiji, Israel, China, and throughout California. At the three-day event, local and international scientists, policy makers, elected officials, corporate sustainability leaders, engineers, planners, natural resource managers, and other practitioners discussed cross-cutting climate change adaptation strategies to be shared across Mediterranean-climate cities. The structure of the conference was designed to maximize cross-boundary thinking, and discussions were organized around topics including water, biodiversity, governance, the built environment and public health, and energy. One key recommendation from the conference was the idea that integration plays a critical role in adaptation and that Mediterranean climate change stakeholders must work together, sharing knowledge and tools, including spatial analysis tools, across topical boundaries and regional borders.

MC-4 has continued to organize its members and partners around the concept of ecosystem-based adaptation. At the 2014 The Mediterranean City Conference in Athens, one of the primary goals was to help local and regional authorities gain insight into the role of Group on Earth Observations (GEO) services in adapting urban areas to climate change. By incorporating GEO into the conference, the organizers (the European Commission, Eurisy, MC-4, and others) sought to make spatial data and analysis a central part of the discourse about the resilience of the five regions.

In addition to leading the 2014 The Mediterranean City Conference, MC-4 has officially participated in events in partnership with the United Nations Framework Convention on Climate Change (UNFCCC), the United Nations Development Programme (UNDP), the Global Adaptation Network (GAN), and the Los Angeles Regional Collaborative for Climate Action and Sustainability (LARC). In 2015, MC-4, LARC, and Los Angeles County collaborated on an event about creating partnerships to use the most updated spatial data for projects with a regional resilience focus called LARIAC Data for a Resilient Los Angeles Region. (LARIAC, for reference, is the Los Angeles Regional Imagery Acquisition Consortium housed at Los Angeles County). This symposium featured speakers discussing spatial data and its applications across sectors; there were representatives from local and regional government, the business community, and academia. It is clear that spatial analysis will continue to play a critical role in MC-4 dialogues and is a key component of future adaptation successes.

In 2017, MC-4 led an official side event at the UNFCCC's Conference of the Parties (COP) 23 in Bonn, Germany, that highlighted the importance of spatial analysis tools in adapting Mediterranean regions to climate change. The side event, Climate Adaptation and Hazard Mitigation: Resilient Mediterranean Cities and Forests, featured speakers from the California State Senate (US), LARC (US), the California Governor's Office of Planning and Research (US), Canberra Urban and Regional Futures at the University of Canberra (Australia), The Nature Conservancy (US), and The Jewish National Fund (Israel). Spatial analysis played a key role, allowing the professionals on the panel to understand adaptation needs and trends at local and regional scales. For example, the CalEnviroScreen tool uses environmental, health, and socioeconomic data from state and federal government sources to help identify California communities that are on the front lines of pollution and climate change impacts. This tool is an important resource across sectors for regionally appropriate climate change planning and policy making. All of the organizations represented at the COP 23 event in Bonn employ spatial data to assist them with the process of planning for the future of their Mediterranean-climate cities.

Overall, MC-4 is a model for integrating inter- and intraregional resilience. In other words, it supports planning within each region while simultaneously coordinating efforts between regions. The methods that MC-4 members are testing and refining in the five Mediterranean regions can also be applied to other networks in ecosystems such as the tropics, deserts, and mountainous areas. In this sense, MC-4 can be viewed as a prototype with global applications.

CLIMATE CHANGE ADAPTATION AND CITIES

Humanity is becoming an urban species. By 2008, more than half of the world's population were city dwellers (World Bank 2014). Today, approximately 54 percent of all human beings live in urban areas, and it is anticipated that this growth trend will continue. By 2045, it is expected that the world's urban population will surpass 6 billion (UN 2014). Further, while in 1950, 30 percent of the world's population was urban, by 2050, 68 percent of the world's population is projected to be urban (UN 2018). Both megacities (cities with over 10 million inhabitants) and small cities are increasing rapidly in number (UN 2014). Medium- and small-sized cities, those with 500,000 people or fewer, are experiencing particularly high rates of growth and currently house 53 percent of all urban dwellers (UN 2014). Growth of cities means an increase in resource use as well. Urban areas currently consume two-thirds of the world's energy production and create over 70 percent of global greenhouse gas emissions (C40 2012). Cities are important when planning for resilience because more and more people are moving into cities. Therefore, it is essential to support efforts that benefit urban areas.

As cities incorporate growing numbers of migrants and rural populations shrink, technical and social questions of how to create and maintain resilient urban systems will become more pressing. For example, in terms of infrastructure, the built-up area of cities worldwide is projected to triple in size between 2000 and 2030, from 200,000 to 600,000 square kilometers. Additionally, larger cities consume two-thirds of the world's energy and create over 70 percent of global greenhouse gas emissions. Maintaining, adapting, and improving city systems is a persistent challenge. In the case of a city's physical structures, once established, they may remain in place for more than 150 years. Addressing these systems with an eye to climate change is of paramount importance because they impact the daily lives of inhabitants over the long term and dictate residents' quality of life.

SPATIAL ANALYSIS PROJECTS
Los Angeles Tree Canopy Assessment and Prioritization Project

As climate change progresses, human populations will likely need to prepare for ecological changes such as increased heat, more frequent extreme hazardous events, and reduced water supply, as well as the associated public health and economic impacts that will result from these drastic changes in the environment. Urban forestry is one strategy that cities often include in their adaptation plans because trees are a resource that can provide myriad social, economic, and environmental benefits. The scientific literature is replete with evidence of these benefits. For example, research has shown relationships between increased tree canopy and decreased asthma rates in children (Lovasi et al. 2008), increased home values (Wachter 2004), and interception of storm water (Xiao and McPherson 2002).

Many urban sustainability and climate adaptation policies include increasing tree canopy as a goal (table 3.1). Yet they often do not provide direction for how this urban forestry strategy will be implemented. We believe it is important that the decisions on where to plant trees are guided by accurate, current data.

Table 3.1. Examples of urban tree canopy policies

City	Name of policy	Tree canopy goal
Baltimore[1]	Sustainability Plan	Double tree canopy by 2037
Chicago[2]	Climate Action Plan	1 million trees
New York[3]	PlaNYC	1 million trees in 10 years
Philadelphia[4]	Greenworks Plan	30% canopy by 2025
Seattle[5]	Urban Forest Stewardship Plan	30% canopy by 2037

1. www.baltimoresustainability.org/plans/sustainability-plan

2. www.chicagoclimateaction.org/

3. www.nyc.gov/html/planyc

4. https://beta.phila.gov/media/20160419140515/2009-greenworks-vision.pdf

5. www.seattle.gov/trees/management.htm

Following the CURes mission to conduct sound urban ecological science to support resilient communities, in 2014 we launched the Los Angeles Tree Canopy Assessment and Prioritization Project. Collaborators in this effort include the Los Angeles Regional Imagery Acquisition Consortium (LARIAC), the University of Vermont, and SavATree Consulting Group. The project applied methods developed by geospatial analysts at the University of Vermont Spatial Analysis Laboratory and the United States Department of Agriculture (USDA) Forest Service (see www.nrs.fs.fed.us/urban/UTC). Their established urban tree canopy assessment protocol brings together high-resolution spatial imagery and lidar data to produce a high-accuracy land-cover classification map. The first phase of this project covered coastal Los Angeles (figure 3.2; for full report, see O'Neil-Dunne et al. 2015). The second phase will cover the entire county of Los Angeles, once countywide lidar data becomes available.

We acquired publicly available 2009 lidar data from the National Oceanic and Atmospheric Administration (NOAA) as part of the California Coastal Conservancy Coastal Lidar Project, and 2014 imagery was acquired through Loyola Marymount University's membership in LARIAC. These provided the basis for the land-cover classification map as shown in figure 3.3. The seven land-cover categories are Tree Canopy, Grass/Shrub, Bare Earth, Water, Buildings, Roads, and Other Paved Surfaces.

Figure 3.2. Boundaries of Phase 1 of the Los Angeles Tree Canopy Assessment and Prioritization Project.

Map courtesy of University of Vermont Spatial Analysis Laboratory.

Figure 3.3. 2009 lidar data (*left*) and 2014 aerial imagery (*middle*) were combined to produce a high-accuracy seven-category land-cover classification map (*right*).

Map courtesy of University of Vermont Spatial Analysis Laboratory.

The land-cover data was then analyzed to derive maps of existing and possible tree canopy percentages at the parcel level (figure 3.4). First, the Tree Canopy category is used to produce an Existing tree canopy layer. The Possible tree canopy layer is composed of Possible Vegetation, which includes the Grass/Shrub and Bare Earth categories, and Possible Impervious, which is the Other Paved Surfaces category.

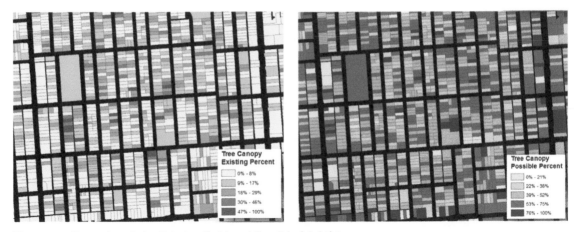

Figure 3.4. Examples of the Existing (*left*) and Possible (*right*) tree canopy maps produced in Phase 1 of the Los Angeles Tree Canopy Assessment and Prioritization Project.

Map courtesy of University of Vermont Spatial Analysis Laboratory.

While the smallest mapped geographic unit was the parcel, the data analysis also aggregated parcels to produce maps of existing and possible tree canopy at the Census Block Group and watershed levels (O'Neil-Dunne et al. 2015). This data can also be aggregated to other useful boundaries, including neighborhoods, cities, and regions. Additional variables of interest, such as surface temperature, demographics, schools, and other critical places can be mapped alongside tree canopy to provide additional information to those tasked with deciding where to plant trees.

The assessment provided the foundation for CURes researchers to begin conducting outreach to determine how this data may be best used to facilitate community prioritization efforts. In other words, how can these high-accuracy maps help guide municipalities and organizations in their tree planting activities? Previous work in New York City (Locke et al. 2011) and Baltimore (Locke et al. 2013) has shown successful application of tree canopy assessments to support decision making. Our aim is to build on those efforts in Los Angeles.

Following the completion of the Phase 1 coastal assessment in 2015, we began a series of community outreach activities to raise awareness of the study and to start to bring together stakeholders around the idea of data-driven tree canopy prioritization. In November 2015, LMU CURes held an event titled Green Priorities in Los Angeles: Data to Support Decision-Making to disseminate the findings of our assessment. About 25 representatives attended this meeting from organizations across the region, including nonprofits, city and county agencies, universities, and public-private partnership organizations. Following this launch event, CURes researchers were invited to speak at various meetings throughout 2016 (table 3.2) to share tree canopy data and promote data-driven tree planting events. This level of community engagement directly supports our mission to conduct applied research in service to our local communities.

Table 3.2. 2016 community events with Los Angeles urban forestry community

Host organization	Description
City of Los Angeles Forestry Advisory Committee	City of Los Angeles council districts are represented, in addition to two seats reserved for nonprofits.
City of Los Angeles Mayor's Office	Representative from several city programs attended, including the resilience officer.
City Plants	Public-private partnership designed to guide the City of Los Angeles' greening efforts.
GreenLA	Coalition of 60+ organizations collaborating to share information and work to improve the environment.
LARIAC	User group for the Los Angeles Regional Imagery Acquisition Consortium (countywide).
LA Urban Center for Natural Resources Sustainability	A partnership between the USDA Forest Service, City of Los Angeles, and several nonprofits.
Los Angeles County	Multiple departments, including Public Health, Public Works, and Regional Planning.
Tree People	Nonprofit serving greater Los Angeles.

These events led to several actions from area organizations. First, the County of Los Angeles decided to fund a land-cover classification project for the entire county in response to a presentation given to the LARIAC user group. This makes the County our partner in Phase 2 of the tree canopy assessment project. Subsequently, we met with representatives from several LA County departments to discuss a joint project to conduct collaborative prioritization workshops in several underserved communities. We are developing a proposal that was scheduled to begin in 2018. We were also able to secure partial funding from Edison International for Phase 2 of the project and received commitment from Tree People to jointly seek funding for the remaining portion of the assessment.

Thus, we used the Phase 1 spatial data to develop new partnerships that provide the foundation needed to support a broader application of the data as we move forward into Phase 2. Our next steps are to use the data to drive prioritization efforts that will lead to increased tree canopy in the Los Angeles region.

Spatial analysis applications for urban animal research

Understanding patterns of movement has been a core feature in the study of animal ecology. Migration, territoriality, and foraging are central themes in research with animal populations. Historically, data was gathered through direct observation and collecting evidence such as tracks, physical remains, or scat piles. New advances in digital spatial analytics are dramatically altering the way in which these questions are being studied and the volumes of data that are gathered.

The emergence of portable radio telemetry in the 1960s was a revolutionary research innovation, in which animals fitted with radio frequency (RF)-emitting devices could be followed and located using directional Yagi-style antennas. Data was typically gathered live by individuals pointing an antenna at a source usually worn as a transmitter by the animal. The data points, gathered in real time, were only as numerous as the amount of effort put in by individual researchers that were following the animal. The data stream was further limited by the broadcasting range of the transmitter, which might be only a few hundred meters, depending on the size of the battery. As a general rule, animals can carry only about 5 percent of their body mass before their behavior is disrupted. So, on an elephant, battery weight is not an issue and the distance for reception might be as much as 5 to 10 miles across an open landscape. For smaller animals, battery size, and therefore transmitting distance, was a significant limiting factor.

This all changed rapidly when remote sensing of animal movement became possible through the use of geopositioning satellites and the ability of radio transmitters to broadcast chunks of data back to waiting researchers. As such, the volume of data has grown by multiple orders of magnitude and provides a level of detail and granularity that was previously impossible. However, the volume of data also provides a significant challenge to researchers as they try to develop systems to sort and aggregate the information in ways that are understandable. Not surprisingly, the enhanced granularity is both a blessing and a curse.

An example from our work in Rhode Island (Mitchell et al. 2015) shows the density of data that can be gathered with respect to native coyotes fitted with radio telemetry collars that store their positional data, which they gather from satellite information, and then broadcast the data to waiting researchers. The collars store the positional information, so the researchers must only make contact with the collars on a weekly basis to download the data. The information transmitted by these collars provides a fine-scale resolution of the behavior patterns and allows researchers to accurately calculate both the defended home range of the animals and the extent of their activity patterns. Figure 3.5 shows data collected from resident and transient coyotes superimposed over eight land-cover classes. Over 30,000 data points were gathered from 31 coyotes. This provided a rich collection of data that allows for individual analysis of the temporal and spatial aspects of the behavior of these animals.

Temporal analysis of the radiotelemetric data showed that individual transient coyotes avoided overlapping with resident individuals. Dispersing and migratory males must be careful not to enter into other, existing coyote territories, which would result in violent confrontations between the dispersing male and other territory-holding adults. This level of resolution provided incredible insight and created opportunities to understand subtle differences in behavior that historically were only possible through direct observation.

While direct observation is always preferable, the labor required to gather that data is nearly impossible and is not easily repeatable. Having fine-scale remotely collected data available in the volumes that are now possible provides a level of analysis that helps wildlife researchers make better decisions about gathering direct observation data and implementing overall management strategies.

THE FUTURE OF SPATIAL ANALYSIS AS A TOOL FOR ECOSYSTEM-BASED CLIMATE CHANGE ADAPTATION

Because Mediterranean regions share similar sets of problems, adaptation solutions are likely to be more easily applicable within these regions than solutions adopted from other climate zones. However, currently there are no mechanisms to encourage Mediterranean cities to share adaptation plans and results. Also, there is little analytical capacity to compare solutions to adaptation challenges across Mediterranean cities.

Mapping provides a powerful tool for building analytical capacity to compare adaptation solutions in the Mediterranean regions, such as urban tree canopy analyses and animal management plans. Further, by using mapping as an analytical tool for this climate zone, we are creating a model that can be replicated in other climate zones around the world. We can also apply this lens at other scales of resilience.

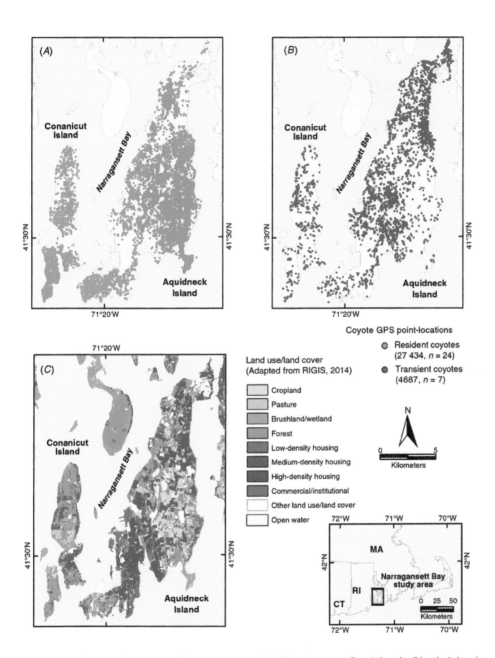

Figure 3.5. The study area for the coyotes on the Narragansett Bay Islands, Rhode Island, shows (A) the distribution of resident coyotes, (B) transient coyotes, and (C) the 10 land-cover classes used in this study (derived from RIGIS 2014).

Map courtesy of Rhode Island Geographic Information System.

For example, we could employ similar tactics for river deltas or other features that haven't traditionally been used as a focal point for climate change adaptation.

The increasing popularity of city-to-city networks, such as C40 Cities Climate Leadership Group, the Rockefeller Foundation's 100 Resilient Cities, and MC-4, demonstrates the rapidly growing need to develop more methods for comparing geographies with similar climate and spatial and socioeconomic characteristics. It is clear that cities are sources of opportunity and innovation for climate change adaptation and that they also serve as nodes for regional capacity to tackle systemwide climate issues. Consequently, effective regional solutions will allow cities to not only accommodate and address social and technical issues but also move to a state of resilience in which they become agile, connected, integrated, and, overall, stronger.

REFERENCES

C40 Cities. 2012. "Why Cities? Ending Climate Change Begins in the City." Accessed September 5, 2017. www.c40.org/ending-climate-change-begins-in-the-city.

Locke, D. H., J. M. Grove, M. Galvin, J. P. O'Neil-Dunne, and C. Murphy. 2013. "Applications of Urban Tree Canopy Assessment and Prioritization Tools: Supporting Collaborative Decision-Making to Achieve Urban Sustainability Goals." *Cities and the Environment.* 6 (1): article 7.

Locke, D. H., J. M. Grove, J. W. Lu, A. Troy, J. P. O'Neil-Dunne, and B. D. Beck. 2011. "Prioritizing Preferable Locations for Increasing Urban Tree Canopy in New York City." *Cities and the Environment* (CATE), 3 (1): 4.

Lovasi, G. S., J. W. Quinn, K. M. Neckerman, M. S. Perzanowski, and A. Rundle. 2008. "Children Living in Areas with More Street Trees Have Lower Asthma Prevalence." *Journal of Epidemiology and Community Health* 62: 647–49.

Mitchell, Numi, Michael W. Strohbach, Ralph Pratt, Wendy C. Finn, and Eric G. Strauss. 2015. "Space Use by Resident and Transient Coyotes in an Urban–Rural Landscape Mosaic." *Wildlife Research* 42: 461–69.

O'Neil-Dunne, J. P. M., D. H. Locke, M. F. Galvin, and T. Engel. 2015. "Tree Canopy Assessment: Los Angeles Coastal Zone." SavATree Consulting Group, Bedford Hills, NY. Accessed June 20, 2016. http://cures.lmu.edu/wp-content/uploads/2015/12 /Tree-Canopy-Report-Los-Angeles.pdf.

Perry, Mark J. 2016. "Economic Output: If States Were Countries, California Would Be France." *Newsweek*, September 5, 2016. www.newsweek.com/economic-output-if-states-were-countries-california-would-be-france-467614.

Rhode Island Geographic Information System (RIGIS) Data Distribution System. Environmental Data Center, University of Rhode Island, Kingston, Rhode Island. Accessed October 9, 2014. www.edc.uri.edu/rigis.

RIGIS. 2014. Land Cover/Land Use for Rhode Island 2011.United Nations. 2018. "World Urbanization Prospects: The 2018 Revision." Accessed June 26, 2018. https://esa.un.org/unpd/wup/Publications/Files/WUP2018-KeyFacts.pdf.

United Nations. 2014. "World's Population Increasingly Urban with More than Half Living in Urban Areas." Accessed September 5, 2017. www.un.org/development/desa/en/news/population/world-urbanization-prospects.html.

Wachter, S. 2004. "The Determinants of Neighborhood Transformation in Philadelphia - Identification and Analysis: The New Kensington Pilot Study." Philadelphia: University of Pennsylvania, Wharton School.

World Bank. 2014. United Nations Population Division. World Urbanization Prospects: 2014 Revision. "Urban Population (% of Total Population)." https://data.worldbank.org/indicator/SP.URB.TOTL.IN.ZS

Xiao, Q., and E. G. McPherson. 2002. "Rainfall Interception by Santa Monica's Municipal Urban Forest." *Urban ecosystems*, 6 (4): 291–302.

→ CHAPTER 4

The mouth of the Columbia River: USACE, GIS, and resilience in a dynamic coastal system

PAUL T. CEDFELDT, JACOB A. WATTS, HANS R. MORITZ, AND HEIDI P. MORITZ

INTRODUCTION

The rising sun lights up the sky near the mouth of the Columbia River. Seagulls screech while double-crested cormorants and Caspian terns dive into the river searching for a meal. The barking of distant sea lions mixes with the rustling sounds of wind and waves. Salmon are running. Many small boats dot the surface of the river, filled with eager sport fishers hoping for a catch. Bells ring and lights blink on the large navigation buoys as an enormous freighter cuts across the water and winds its way up the estuary (figure 4.1). The freighter has just completed a journey across the Pacific Ocean to the mouth of the Columbia River, threaded its way between the jetties and across the bar, and finally arrived in the Columbia River estuary. Although its journey is coming to an end, it isn't over just yet. The ship will continue to make its way up the estuary for several days, following the navigation channel until it reaches a port more than 100 miles upriver and fills its hold with wheat before turning around. It will then retrace its path, winding back down the

The views, opinions, and conclusions expressed herein are those of the authors and do not necessarily reflect the official position of the US Army or the Corps of Engineers.

mighty Columbia, going again across the bar and past the jetties at the mouth, and finally returning to the wide expanse of the Pacific Ocean.

Figure 4.1. Sunrise is shown at the mouth of the Columbia River.

Photo courtesy of Jacob Watts.

In this chapter, we tell the story of how GIS is being used to help achieve resilience at the mouth of the Columbia River, first touching on the history of the river and its people, and then shifting our focus to the role of the US Army Corps of Engineers. Multiple, specific examples will be provided, detailing how this federal agency is currently using GIS to help this area achieve resilience.

Place history

The Columbia River is the fourth-largest river in North America as measured by average annual flow, and along with its major tributaries, the Snake and the Willamette, it forms the dominant river system in the Pacific Northwest region of the United States. The Columbia is often referred to as the "Great River of the West," immortalized in the 1940s by the folk singer-songwriter Woody Guthrie in "Roll On, Columbia, Roll On" and his many other songs about the Columbia River (BPA 2012). From its headwaters in British Columbia, Canada, the great river flows over 1,200 miles through Canada and the United States to its outlet into the Pacific Ocean near Astoria, Oregon (Haglund 2011, 4). Depending on the time of year, the Columbia River accounts for between 60 percent and 90 percent of the total freshwater discharged into the Pacific Ocean from the Canadian border south to San Francisco, California. The Columbia River Basin includes 3,000 square miles of waterways and lakes. The river, which serves as a large portion of the Washington and Oregon

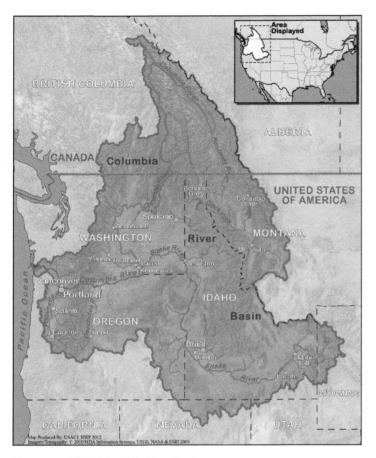

Figure 4.2. The Columbia River Basin.

Map created by Jacob Watts. Data courtesy of USACE, MDA Information Systems, Bureau of Transportation Statistics NTAD 2004, USGS/EPA, NASA, Canadian GeoBase and Esri.

state boundaries, drains an area of approximately 219,000 square miles in the states of Washington, Oregon, Idaho, Montana, Wyoming, Nevada, and Utah. An additional 39,500-square-mile portion of the basin lies within Canada. To give some perspective, the Columbia River Basin's 258,500 square miles make it similar in size to the country of France (figure 4.2).

The people

There is evidence of human habitation in the Columbia River Basin dating back over 14,000 years. These Native Americans were the first human residents in the area, and their descendants continue to live throughout the basin today. They relied on hunting, gathering, and most importantly, fishing. The Columbia's abundant salmon were particularly important to the survival, culture, and spiritual life of these

Native Americans. In the Sahaptin language, the Columbia was the "Great River," or "Nch'i-Wana." The Columbia and its salmon continue to be an essential part of contemporary Native American culture. Figure 4.3 shows the current reservations of federally recognized Native American tribes in the region. Tribes with treaty rights and still fishing the Nch'i-Wana include the Confederated Tribes and Bands of the Yakama Nation, the Confederated Tribes of the Umatilla Indian Reservation, the Nez Perce Tribe, and the Confederated Tribes of Warm Springs. Although their reservations may not be adjacent to the Columbia, their tribal members are guaranteed access to the river and to fish year-round.

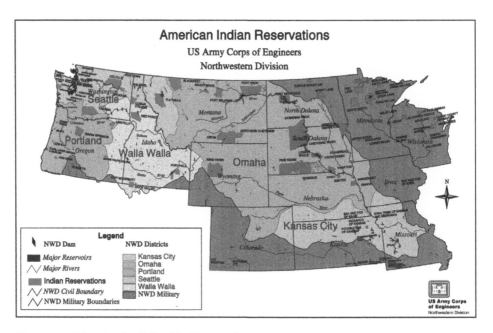

Figure 4.3. Tribes in the Columbia River region.

Initial contact with European fur trappers and traders, and associated viruses—most notably smallpox—decimated Indian populations in the late 1700s. White settlement of the area was opened up after the Lewis and Clark expedition from 1804 to 1806, but settlers did not arrive en masse until the late 1840s. Farming, timber harvesting, and fishing were the predominant drivers of the economy throughout the Columbia River Basin. At the mouth of the Columbia River, commercial fishing and the associated canneries became major industries beginning in the 1860s. While important to the settlement of the region through the 1800s and 1900s, logging and fishing are no longer the primary economic drivers in the area. Timber harvesting had dropped off by the early 1990s. The last commercial cannery operated by Bumble Bee Seafoods in Astoria, Oregon, closed in 1980. Today, there are three ports near the mouth of the Columbia River. On the Washington side, the ports of Ilwaco and Chinook support commercial and recreational fishing, and seafood processing. The larger Port of Astoria in Oregon supports cruise ship calls, log exports, and seafood processing.

In Astoria, the current population is approximately 10,000. According to the 2010 census, more than 90 percent of residents are white (US Census Bureau 2016). Many current residents in the area are of Scandinavian descent, and there is still an annual Astoria Scandinavian Midsummer Festival. Astoria also hosts a Crab, Seafood & Wine Festival every April, and the Astoria International Film Festival in October. The Columbia River Maritime Museum in Astoria is the official state maritime museum of Oregon. Tourism, including the hosting of cruise ships, is now a major economic driver. Fishing and logging continue to be an important but smaller component of the local economy.

US Army Corps of Engineers and the mouth of the Columbia River

Over time, the outlet at the mouth of the Columbia River has developed an infamous reputation. The large number of shipwrecks (figure 4.4) in the bar area, almost 2,000 since 1792, has earned it the nickname "Graveyard of the Pacific" (Haglund 2011, 12).

The Columbia River Bar is approximately 6 miles long and 3 miles wide. It encompasses the area between the north and south jetties and the area from the mouth of the river to Sand Island. In his history of the Columbia River Bar, Michael Haglund (2011) convincingly argues that the bar is the world's most dangerous shipping passage. This is primarily due to the unique geologic history of the area. Starting approximately 15,000 years ago, the massive Missoula floods carved out the Columbia River Gorge and formed the relatively narrow outlet responsible for extreme wave action and currents (Haglund 2011, 3–4). During winter storms, waves can be as high as 30 feet at the Columbia River Bar. These large waves combine with strong outgoing tides to create extremely hazardous conditions. The US Coast Guard station at Cape Disappointment, located near the mouth of the Columbia, conducts an average of 450 to 500 search and rescue cases every year (Coast Guard website 2020). The Cape Disappointment station was featured in 2014 for a full season of the Weather Channel's TV show featuring Coast Guard operations in

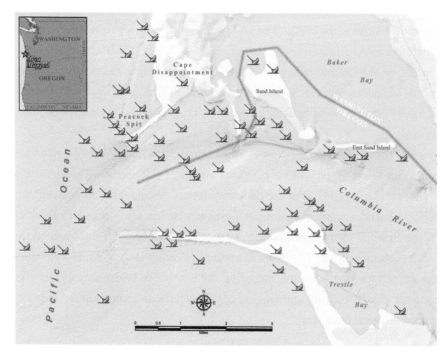

Figure 4.4. The "Graveyard of the Pacific" shipwreck locations, 1850–1987.

Map created by Jason Miller, USACE.

Figure 4.5. Photo of the 209th Engineers, Camp Sheridan, Alabama, 1919, forming the Corps Castle. This insignia is associated with the construction of coastal defense fortifications, one of the earliest Corps responsibilities.

Photo courtesy of the USACE Office of History.

extreme environments, identifying the Columbia River Bar as one of the most hazardous environments in North America.

For more than a century, the United States Army Corps of Engineers (USACE) has been involved with helping to manage the hazardous conditions at the mouth of the Columbia River (MCR). Army engineer officers were first appointed in 1775, but the Corps of Engineers was officially established as a separate Army entity in 1802. USACE was initially responsible for founding and operating of the military academy at West Point (figure 4.5) (Army Corps of Engineers Office of History 2007).

Early on, USACE was directed to improve waterway navigation for the dual purpose of facilitating the movement of the United States Army and its materials and contributing to national economic development by, in the language of the time, "improving" waterways to ensure the safe movement of goods and people. USACE has maintained this navigation mission and has over the years expanded its Civil Works mission focus to also encompass hydropower, flood risk management, recreation, and environmental restoration. For over 200 years, USACE as an agency has demonstrated its resilience from both an organizational perspective and through its efforts constructing projects that have helped many communities to maintain resilience in various parts of the country. When faced with changing national priorities, the agency has flourished by transforming itself to accommodate multiple new missions and challenges. In *Staking Out the Terrain: Power and Performance among Natural Resource Agencies,* the authors, Jeanne Nienaber Clarke and Daniel C. McCool (1996), identify USACE as a bureaucratic superstar: "The Corps rarely turns down an opportunity to expand its areas of responsibility. It even takes challenges to its developmental orientation as opportunities to demonstrate its responsiveness to changing public values" (figure 4.6) (Clarke and McCool 1996, 23).

US Army Corps of Engineers and the Pacific Northwest

USACE has been an influential presence in the Pacific Northwest since 1871, when the Corps' Portland Engineers Office, the forerunner to the Portland District, was established in Portland, Oregon, by Major Henry M. Robert. Since its inception, the Portland District has focused on managing the Columbia River and its major tributaries for navigation. Although the Portland District mission has expanded over time to include additional USACE Civil Works missions, ensuring navigation on the Columbia River has been the common thread throughout its history. The Army Corps of Engineers has long incorporated resilience into its projects and delivered solutions at the MCR that enhance resilience at both local and regional scales.

Figure 4.6. The *Benjamin Humphreys*, a combination snag boat and bucket dredge, operating on the Mississippi River in 1908.

Photo courtesy of the USACE Office of History.

Figure 4.7. The mouth of the Columbia River is shown as it empties into the Pacific Ocean.

Photo courtesy of USACE Portland District.

The Portland District's direct involvement with managing the mouth of the Columbia River dates to the Rivers and Harbors Act of 1884 (USACE Portland District 2012). This was the authorizing legislation for the MCR navigation project and the three primary navigation structures: the jetties. The jetty system consists of three rubble-mound jetties: South Jetty, North Jetty, and Jetty A (figures 4.7 and 4.8). These jetties were constructed in different phases from 1885 to 1939 to secure consistent navigation through the coastal inlet. Their sequential design and construction is a good example of adapting and working within the mouth of the Columbia River system.

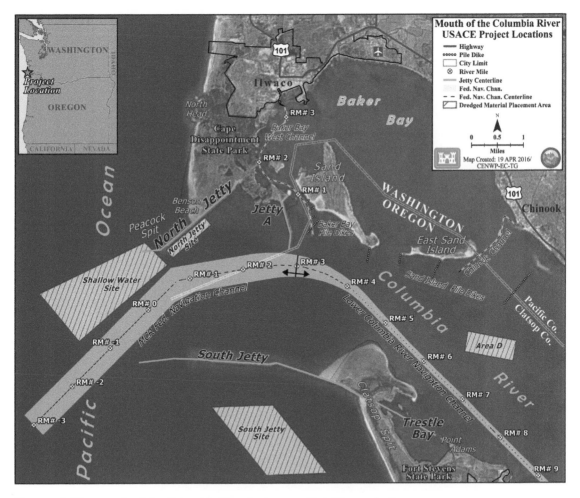

Figure 4.8. The mouth of the Columbia River is shown with jetties and other USACE projects.

Map created by Jacob Watts, USACE.

Due to the magnitude and cost of jetty construction at the mouth of the Columbia River, it was conducted in stages, with each successive stage only beginning once the inlet and navigation channel response to the previous construction was observed. The South Jetty is located on the Oregon side of the river. The initial 4.5-mile section of the South Jetty was completed in 1895, with a 2.5-mile extension completed in 1913. The North Jetty and Jetty A are located on the Washington side. The North Jetty was completed in 1917. Jetty A, positioned near the root of the North Jetty, was constructed in 1939 to a length of 1.1 miles. Jetty A was intended to direct river and tidal currents away from the North Jetty foundation. Jetty construction changed the

Figure 4.9. North Jetty construction in 1916.

Photo courtesy of the *Chinook Observer*.

Figure 4.10. South Jetty repair in 1936.

Photo courtesy of USACE Portland District.

ocean entrance to the Columbia River, supports a consistent navigation channel across the bar, and significantly improved navigation through the MCR. In this dynamic coastal environment, the morphology of the inlet constantly changes. The jetties are constantly deteriorating as they are continually exposed to extreme wave attack and erosion of the tidal shoals on which they were built (figures 4.9 and 4.10).

Local importance of the jetty

Multiple stakeholders, including over 100 ports, businesses, and other organizations represented by the Pacific Northwest Waterways Association (PNWA), rely on a functioning jetty system as a foundation for the economic resilience of national, regional, and local economies (PNWA 2016). According to the Center for Economic Development and Research (2005), the Columbia/Snake River navigation system is the nation's number one export gateway for wheat and barley exports. It is also the number one export gateway for West Coast wood, mineral bulk exports, and automobile imports. Functioning jetties at the mouth of the Columbia River annually support $20 billion in international trade, 42 million tons of cargo, 3,500 cargo vessel crossings, and more than 40,000 jobs dependent on this trade. On a more local level, both commercial and recreational fishing crews and crabbers rely on the protection of the jetties to ensure safe passage of their vessels. The Northwest Sportfishing Industry Association (NSIA) is another stakeholder in the success of USACE efforts at the mouth of the Columbia River. This lobby group represents the business interests of members that depend on the sport of fishing for their livelihoods.

In the years since the jetties were constructed, significant funding and engineering effort has been directed by the Portland District to maintain the Columbia River navigation system via infrastructure design inclusive of resilience. The USACE motto is "Building Strong," and part of this is building resiliency into infrastructure repair. For this project, resiliency also refers to understanding how the structures interact with the hydrodynamics of the inlet and the underwater shoals. USACE has a history of adapting the latest technology in pursuit of engineering excellence. GIS is one of the most recent technological tools being incorporated into the process of maintaining navigation at the MCR. For example, planning and design for rehabilitation of the jetties at the MCR make extensive use of GIS for spatial data management, analysis, and information dissemination.

In 2016, the North Jetty rehabilitation project required a volume analysis of the quantity of stone required to reconstruct the North Jetty of the Columbia River to a full design template. These volume calculations are used by USACE coastal engineers to assess multiple template design alternatives. Additionally, a temporal change analysis is performed by calculating volume differentials between elevation surfaces derived from surveys performed intermittently over the previous decade. These calculations are used to identify the rate of degradation along the jetty over time. To analyze the temporal and design volumes relative to past repair locations, in conjunction with the spatial variability of rock movement due to structural and environmental factors, the analysis is conducted in 50-foot bins along each side of the jetty centerline. The Portland District team of geospatial professionals conducts the volumetric analysis, starting with data acquisition and management, then GIS analysis, and finally cartographic and tabular distribution of the output information to the rest of the technical team. Specific steps of this process and tools used are described in the following section.

TECHNICAL ANALYSIS: NORTH JETTY REHAB PROJECT

Introduction

The following technical write-up describes how the Portland District (NWP) geospatial professionals engage in important geospatial analyses in support of the North Jetty Rehab project, from contracting for high-resolution remote sensing-based surveys, to manual and automated geoprocessing, and then to tabular and cartographic information dissemination to the product delivery team (PDT). The processes and products described here are used on numerous NWP jetty projects and can be used by other coastal engineering entities doing similar work.

Survey data acquisition/data management

The primary input to the volumetric analysis at the MCR North Jetty is a survey of the current elevations on and around the jetty. To obtain this data, NWP contracts high-resolution topographic and bathymetric surveys. Using the district's established remote sensing indefinite delivery, indefinite quantity (IDIQ) contract mechanism, a geospatial team member serves as technical representative on the task order and guides the survey from scope of work (SOW) and independent government estimate (IGE) development to delivery of a photogrammetric topographic survey and a multibeam hydrographic survey, which are combined into an integrated topographic/bathymetric digital elevation model (topo/bathy DEM) of the jetty (figure 4.11).

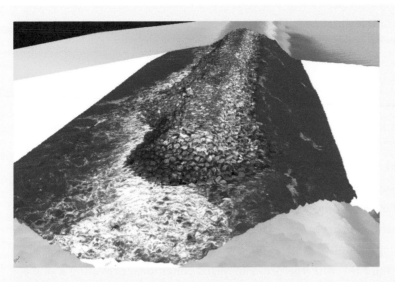

Figure 4.11. Example of integrated topo/bathy terrain with partial imagery overlay.

Image created by Jacob Watts, USACE.

The photogrammetric survey is performed by a licensed professional photo-grammetry firm. An automated photogrammetric processing technique supplemented with traditional photogrammetry is currently used. High-resolution (5-cm pixels) aerial photography is collected along the length of the jetty at low tide. The imagery is run through an automated procedure that generates an elevation point cloud (400 x,y,z points per square meter) and orthorectified imagery. The automated process generates a digital surface model (DSM), which maps elevations for everything observed in the imagery. In areas in which there is dense vegetation or sea lion colonies, traditional photogrammetry is used to produce elevation break-lines that allow for a bare-earth representation in the final DEM. The raw survey deliverable is then processed by the NWP geospatial team into 3D multipoint and polyline feature classes in ArcGIS and serves as the topographic input into the final integrated DEM.

As noted, the mouth of the Columbia River is a volatile location that can make hydrographic surveys difficult to execute. When conditions are deemed acceptable from a safety and data acquisition standpoint, the contractor deploys a team of highly trained hydrographic surveyors on-site in a survey vessel equipped with survey-grade GPS equipment, multibeam sonar sensors, and data acquisition computer hardware and software. The acquired data is then calibrated and edited. A gridded (equally spaced) x,y,z point cloud is delivered as a .txt file, which is then processed into a 3D multipoint feature class. This serves as the bathymetric input into the final integrated DEM.

GIS analyses

With the surveys processed into GIS data, the analyst then uses them to generate a terrain dataset in ArcGIS. The terrain dataset allows for the 3D multipoint and polyline feature classes to be combined and interpolated into a dynamic triangulated irregular network (TIN). For use in the ensuing analyses, the terrain dataset is converted into a raster dataset, using a cell resolution that retains the integrity of the highly detailed survey (0.5-foot cell size) in the DEM.

The orthorectified imagery also plays a critical role in providing visual information of the existing conditions along the jetty. The current high-resolution imagery is compared by the PDT with the previous year's conditions, providing a powerful visual analytical tool for detecting and monitoring change to the structure. The contractor delivers the imagery as tiled orthorectified images, and the NWP analyst mosaics the tiles into a single georeferenced .img file, which is then imported into the project geodatabase.

With the survey data and orthoimagery processed, the analyst then performs quality assurance by reviewing the output datasets to verify that the survey meets the specifications set forth in the scope of work. To do this, the survey data is compared against previous years' DEMs and known elevation points. A spot check is appropriate to verify whether the data at known locations falls within the tolerance level of the specifications and that coverage is acceptable. The imagery is reviewed to ensure full coverage, appropriate lighting, and that all required bands (RGB and near-infrared) were delivered.

Temporal and design condition surface differential and volumetric quantities for entire jetty

An initial temporal analysis of the entire jetty is performed for the purpose of demonstrating the year-over-year change along the jetty. Two separate analyses take place to quantify the total volume of rock that has been added or removed since the previous survey. A surface differential calculation demonstrates the difference in elevation at specific locations from previous surveys to the current one (figure 4.12).

The Cut Fill tool in ArcGIS is used to calculate the change in the volume of rock along the jetty. Using the current year and historical DEMs, the analyst generates both a raster file and a tabular attribute table that show the locations where

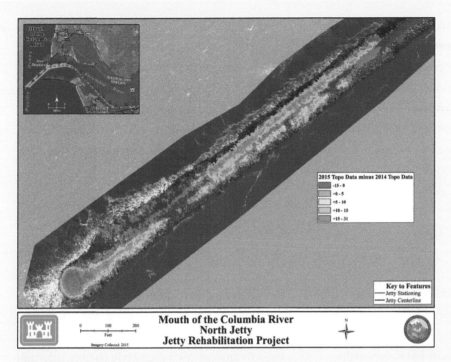

2015 Topo Data minus 2014 Topo Data
- -15 - 0
- +0 - 5
- +5 - 10
- +10 - 15
- +15 - 31

Key to Features
— Jetty Stationing
— Jetty Centerline

Mouth of the Columbia River
North Jetty
Jetty Rehabilitation Project

0   100   200
Feet
Imagery Collected: 2015

N

Figure 4.12. 2014 versus 2015 surface differential map.

Map created by Jacob Watts, USACE.

the surface elevations have changed from year to year. The raster is symbolized by areas of cut, fill, or no change. The output cut fill attribute table is summarized to describe the total amount of change between the existing condition of the jetty and that of previous years. In most cases, the jetties only degrade over time, but in years where repair work has been completed, this analysis also demonstrates areas where fill activity occurred.

Where the Cut Fill tool provides the volume of change that has occurred year over year, the Minus tool is used to generate a surface differential raster dataset that retains the elevation change between the existing condition and previous years on a cell-by-cell basis. The output cell values are either positive or negative numbers that demonstrate, in feet, the difference between the two surveys at that location. The output raster can be symbolized and mapped in ways that allow users to quickly analyze the degree of degradation at a given location along the jetty over the specified period of time.

In addition to the temporal analyses, the PDT needs to understand how the existing condition of the survey relates to alternative jetty rehabilitation design templates. The template designs are provided to the analyst by district coastal engineers in the form of a textual description as follows: design template stationing, crest width, crest elevation, and side slope. The analyst uses the specifications to generate a raster dataset of the 3D design template that is analyzed against the current DEM to demonstrate the volumetric quantities and surface differential between the template and existing condition. The Cut Fill and Minus tools described earlier are also used to generate full jetty volume computations and surface differential data between the existing and design conditions.

Temporal jetty design template versus existing condition volumetric quantities in 50-foot analytical bins to produce tabular data output for subsections of the jetty

District coastal engineers require a tabular spreadsheet demonstrating the volumetric differential between existing and design conditions at specified intervals along the jetty. The output of this process does not produce a map product but generates a Microsoft Excel spreadsheet that is used to analyze the volumetric change information in subunits along the jetty. For this analysis, a 50-foot analysis bin dataset is created, and then an ArcGIS ModelBuilder™ geoprocessing routine is built to automate the statistical geoprocessing procedure (figure 4.13). For this purpose, the term *analysis bin* is used to describe a polygon feature class that contains rectangular features that are used as subareas for the purpose of summarizing volume computations in subsections of the jetty. The polygons are created in 50-foot increments along the jetty centerline, extending horizontally from the centerline on both sides of the jetty for a distance that captures the entire area of interest for the volume analysis (100 feet from centerline in this case).

ModelBuilder is then used to automate the process for generating tabular volumetric data demonstrating the difference between the jetty template design and the existing condition. The process is run over DEMs from multiple years, and the output exhibits how the jetty has changed over time in the smaller analytical areas. This provides the information needed to analyze degradation in specific sections of the jetty, as the structure's size introduces varying dynamics in play along the jetty that affect it differently from location to location. The geoprocessing model allows the analyst to replicate the process over multiple years in a standardized fashion.

The following steps outline the geoprocessing tools that are required to generate the tabular volumetric data. ModelBuilder is used to string the operations together

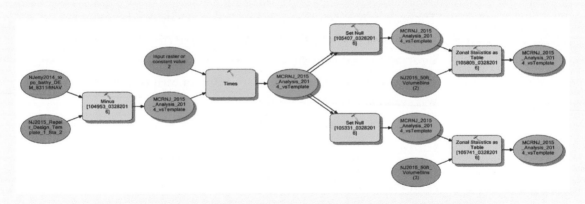

Figure 4.13. Geoprocessing model for tabular volumetric analysis.

into a single tool that maintains consistency in parameter selection throughout the analysis.

1. The Minus tool is the first step in the process, run between the jetty template design and the DEM for the year's analysis. This gives the elevation difference between the two datasets.

2. The Times tool is then used to calculate the differential volume on a cell-by-cell basis. For the MCR North Jetty analysis, the DEMs were produced using a 0.5-foot cell size. Thus, the pixel's volume calculation is as follows: $(0.5 \times 0.5) \times \Delta z$. The output raster dataset represents the volume of existing material above and below the design template on a cell-by-cell basis.

3. The analyst separates the data into the volume of material above and below the jetty template design. To do this, the Set Null tool is run two times on the output raster from the Times tool—once with an expression written that will create a new raster with just those values that are greater than zero and a second time with values that are less than zero (or above/below the design template).

4. The final step of the geoprocessing routine is to run the Zonal Statistics as Table tool twice, once on the above template raster generated with the Set Null tool and once on the below template raster. The 50-foot analysis bin feature class provides the geometry and naming data for this step, and the output rasters from the two iterations of the Set Null tool will serve as the input volumetric data. This generates an output table that retains the values for the sum of all the volumetric differential within each 50-foot bin.

5. The analyst then runs the output table through the Table to Excel tool, producing tabular data that is loaded into a primary spreadsheet retaining separate columns with the volumetric quantities, by 50-foot analysis bin, above and below the jetty template design, for each year that survey data was available.

Conclusion

GIS has been shown to be a powerful tool for analyzing the rate of degradation to jetty structures as well as aiding in alternative design selection. This write-up has discussed some of the ways NWP's team of geospatial professionals contribute by guiding the process from survey data acquisition to engineering design decisions and construction plans. The processes and products described here are transferable to various other project types and can be employed by other coastal engineering entities that require similar products.

Accurate and up-to-date nautical charts are an essential component of successful marine navigation. This is another area in which GIS is extensively used by USACE to deliver an essential product relied on by sailors for safe passage through the Columbia River system. Many USACE districts routinely collect hydrographic survey data to assess channel depths, determine the need for and progress of dredging, and inform other USACE offices and navigation interests of channel conditions. A new GIS-based system for managing hydrographic survey data and producing charts was pioneered at the Portland District and then adapted Corps-wide as the eHydro program. Before eHydro, the process of converting channel survey data into nautical charts was handled differently by each district. At the Portland District, the older method used a CAD system to plot the survey depth soundings. These soundings were then visually reviewed to find any areas of shoaling to inform future dredging efforts. This process was subjective, in that it relied on the best professional judgment of a technician reviewing the chart. It could also take weeks to produce a single chart. GIS, via the eHydro application, allows USACE to collect and process channel survey data more objectively and introduces standards to make the process repeatable and consistent.

The eHydro application is a set of custom tools that run within ArcGIS. Python scripting is used extensively to do the initial processing of survey depth soundings. The raw input data is a text file of depths in x,y,z format. Python scripts automate the x,y,z data geoprocessing using ArcGIS tools. The scripts output sounding points, contour lines, and shoaling areas.

Python is then used to manipulate a chart template and export a map to a PDF file. Charts are now much more quickly available for review. It typically takes only 10 to 15 minutes to produce a chart, whereas before eHydro, the manual process took many hours. eHydro data is managed in a standardized geodatabase that enforces naming conventions for layers and x,y,z input files (figure 4.14). This provides a location for all data layers in a single, consistent data repository and allows easy integration of new data.

Data Management District Level

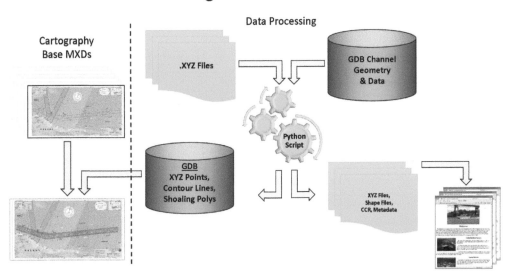

Figure 4.14. eHydro district-level data management.

After outputs are reviewed and approved, eHydro automatically exports channel condition data products to a national, enterprise data system. The enterprise system uses web services to export soundings and contours so that stakeholders and users outside of USACE, such as the National Oceanic and Atmospheric Administration (NOAA), can analyze the data and develop their own information products (figure 4.15).

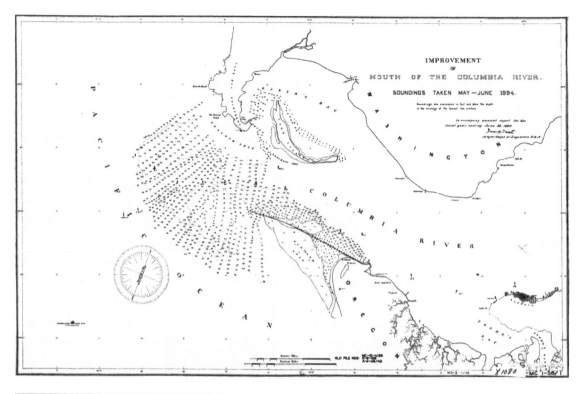

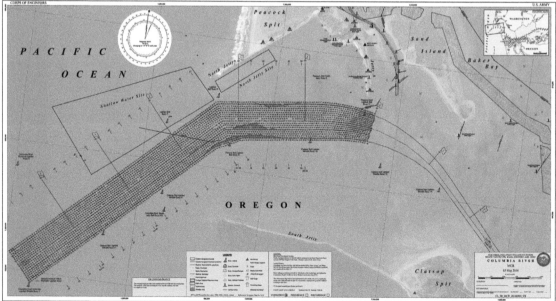

Figure 4.15. Nautical charts, 1894 and 2016.

Environmental movement and USACE

All of USACE's missions, including navigation, were significantly affected by the environmental movement of the late 20th century. However, this provided an opportunity for USACE to once again demonstrate its organizational resilience. Changing national attitudes on the environment began to be reflected in national policy, culminating in the National Environmental Policy Act (NEPA) of 1970. Again, from Clarke and McCool: "The Corps' response to the environmental movement of the 1960s and 1970s is further demonstration of the agency's flexible, even innovative, style. Several analyses of the Corps' response to NEPA generally reach the same conclusion: The agency's response was sincere, swift, and impressive" (1996, 23). The environmental movement, and the resulting change in USACE missions to include environmental restoration, has profoundly influenced the Portland District. In 2001, Davis Moriuchi, the Portland District's former deputy district engineer for project management, explained, "There is a heightened environmental ethic here [in Portland] and has been for some time; it has maybe put us in the forefront in dealing with environmental issues" (USACE Portland District 2003, 81). GIS has again proven to be an indispensable tool for engineering—in this case, for pursuing environmental restoration. At Trestle Bay, GIS is helping the Portland District come full circle, taking responsibility for and helping mitigate some of the environmental damage done while constructing infrastructure at the mouth of the Columbia River.

Trestle Bay is one of the sites that the Portland District identified as a candidate for ecosystem restoration actions under the authority of Section 536 of the Water Resources Development Act of 2000. Section 536 specifically authorizes restoration activities in the lower Columbia River. The Trestle Bay site is in Oregon, near the mouth, at approximately river mile six of the Federal Navigation Channel. At Trestle Bay, approximately 8,800 linear feet of permeable rock jetty bisects the bay and encloses 628 acres of intertidal and shallow subtidal habitat. This is the remnant of the original South Jetty construction started in 1885.

Trestle Bay now provides critical habitat and foraging areas to a variety of fish and shellfish species. Salmon access to the bay was limited by the jetty remnant that bisected the bay. The purpose of the Trestle Bay project was to improve access, opportunity, and ecological function for the benefit of juvenile salmonids and other estuarine fish species. The project was intended to provide improved foraging and rearing conditions for juvenile salmonids by increasing access and egress to the bay's important shallow-water habitat. Improved access, opportunity, and ecological function are expected to benefit several threatened and endangered species, including fall and spring/summer Chinook salmon, chum salmon, Snake River sockeye salmon, steelhead trout, Coho salmon, and coastal cutthroat trout.

GIS for environmental restoration

As described earlier, the need for the Trestle Bay project was predicated on the significant wetland and tidal estuarine habitat losses that have occurred near MCR and within the larger lower Columbia River estuary. In 2016, USACE produced plans and specifications to create several notches in the remnant jetty to improve access and egress to the bay throughout the entire tidal cycle. GIS was instrumental to this task and was used extensively throughout the life of the project. In a similar fashion to the North Jetty GIS analyses described in the technical write-up earlier, the Portland District team of geospatial professionals contributed to the Trestle Bay project throughout the study, design, and construction phases. From contracting for surveys to building an integrated topo/bathy DEM of the existing condition, the GIS team provided the baseline of information that would facilitate hydraulic modeling for alternative design selection as well as volume computations of the alternative design concepts. GIS also played an important role in disseminating pertinent information to both internal and external stakeholders by producing two-dimensional and three-dimensional cartographic products demonstrating the project design features and how the alternative designs would interact with the existing conditions at the site (figure 4.16). Upon selection of the final design criteria, the geospatial team worked closely with CAD technicians to produce the construction plans for the Plans and Specifications package that was used by the construction contractor to build the project as required. This work included design specifications and volumetric quantities for the notches where rock was removed. Cross-section data was also generated in the GIS to demonstrate the design condition in a standardized graph format.

GIS has proven to be a powerful tool that helps USACE meet its environmental restoration mission. This important work is contributing to environmental resilience in the lower Columbia River by improving and expanding salmon habitat. In *Salmon Without Rivers: A History of the Pacific Salmon Crisis,* James Lichatowich argues that restoring natural pathways through healthy riverscapes is the key to restoring salmon. At Trestle Bay, USACE is directly contributing to this type of solution while also aiding the resilience of the tribes in the Columbia River Basin. For thousands of years, Native Americans in the Columbia River Basin have used salmon as an essential resource. Their culture and economy developed around the use of this resource. The Columbia River Inter-Tribal Fish Commission (CRITFC) coordinates

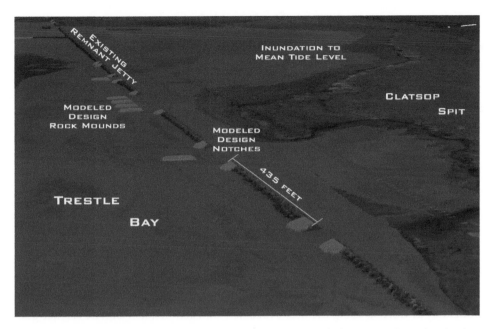

Figure 4.16. Trestle Bay. Digital representation of project, including several notches in the remnant trestle.

management policy and provides fisheries technical services for the Confederated Tribes and Bands of the Yakama Nation, the Confederated Tribes of Warm Springs, the Confederated Tribes of the Umatilla, and the Nez Perce Tribe. CRITFC identifies the Tribal Salmon Culture in the Pacific Northwest as completely reliant on salmon as an integral part of tribal religion, culture, and physical sustenance. Salmon is at the heart of these tribes' cultural resilience. According to CRITFC, "Without salmon returning to our rivers and streams, we would cease to be Indian people" (CRITFC website, last accessed 2020). USACE has a long history of creating dam infrastructure that has made it challenging for tribes to maintain their connection to the Columbia River and exercise their right to subsistence fishing. Trestle Bay represents a small step toward mitigating this challenge and aiding tribal resilience through the restoration of salmon habitat, and GIS is playing an important role in studying, designing, and implementing engineering solutions.

USACE, GIS, AND RESILIENCY

Looking further into the future of USACE and the continued use of GIS for resiliency, it is now the official policy of the Corps to integrate resilience planning and actions into its missions, operations, programs, and projects. The organization's Resilience Initiative Roadmap for 2016 articulates the USACE approach to resilience in accordance with President Barack Obama's Executive Order 13653, *Preparing the United States for the Impacts of Climate Change*, which defines resilience as "… the ability to anticipate, prepare for, and adapt to changing conditions and withstand and recover from disruptions." USACE has developed two primary approaches for implementing resilience strategies, including evolving USACE standards and criteria, and supporting community resilience. As discussed earlier, supporting resilience has already been a part of USACE projects, but this new road map acknowledges there is more that can be done and outlines the way forward for continued improvement. The work to understand and adapt to the effects of climate change is already underway at USACE using the best available—and actionable—climate science and climate change information. Resilience is also an explicit component of the USACE Climate Adaptation Mission, which is "To improve resilience and decrease vulnerability to the effects of climate change and variability."

Coastal risks to resilience

Coastal areas such as the mouth of the Columbia River are especially vulnerable to hazards, now and in the future, posed by waves and surges associated with sea level change and increasingly intense coastal storms. Coastal risk reduction can be achieved through a variety of approaches, including natural or nature-based features (for example, wetlands and dunes), nonstructural interventions (for example, policies, building codes, and emergency response such as early warning and evacuation plans), structural interventions (for example, seawalls and breakwaters), and modifications to projects and project management (for example, added robustness and increased maintenance).

USACE has used GIS to help reduce coastal risks from, and improve resilience to, these hazards. At the national level, GIS was used to map USACE project locations across the country in relation to sea level trends, the coastal vulnerability index, principle ports, and population centers (figure 4.17).

Figure 4.17, a map generated in ArcGIS at the Portland District, represents a collaborative multiagency effort to understand and convey the risks associated with sea level change. The map overlays various key datasets to illustrate the importance

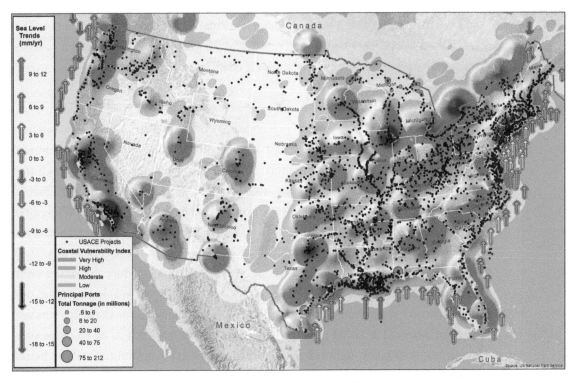

Figure 4.17. USACE projects, sea level trends, and coastal vulnerability.

Map created by Gregg Bertrand, USACE.

of incorporating sea level change science into engineering design and public policy decisions. Data displayed includes NOAA sea level trends as graduated arrow symbols and the US Geological Survey's (USGS) Coastal Vulnerability Index (CVI) values as a multicolored ribbon along the coastline. The CVI analysis used six variables and parameters to demonstrate coastal vulnerability to sea level change and mapped them categorically by color, showing the degree of vulnerability in a given area on a scale of very high to low. The map uses a heat map layer to identify areas of high population density and graduated purple dots to show the location and relative annual tonnage of the nation's ports. It also identifies USACE projects as black dots to put into context the impact of sea level change on USACE infrastructure. The information conveyed struck a chord in the USACE research community and had a large national impact, helping to frame the discussion around USACE's response to sea level change. It clearly shows that inland areas are tied to coastal areas by a complex network, including the inland navigation system, road and rail transport, and supply chains, while also demonstrating the high coastal population densities and the magnitude of the associated economic damages from sea level change. The USGS and USACE are continuing to collaborate on a more sophisticated analysis that will better define the risks of sea level change to Corps projects.

Climate change resilience

Back at the mouth of the Columbia River, the Portland District is already building climate change resilience into its coastal infrastructure. The winter storms that regularly pound the Oregon and southern Washington coasts are much like the hurricanes experienced on the East Coast. According to Heidi Moritz, Portland District coastal engineer, "Our average winter storm wave height is four meters. While challenging, these conditions give the Portland District an advantage when planning for climate change" (Portland District 2015). Rehabilitation of the jetties, as described earlier, incorporated design elements to be resilient to and reduce the future risks associated with climate change. According to Moritz, "If we're able to look at our infrastructure projects in a scenario-based approach, and think ahead for a range of conditions, what could possibly happen and what we need to do to make sure we're prepared for it, then the people in those coastal communities will be better served, better protected, and more resilient" (Moritz 2016). The Portland District ensures the resilience of the mouth of the Columbia River project through a range of interactive elements, including navigation structure monitoring and maintenance, investigating the changes in project morphology (underwater shoals and adjacent shorelines), careful placement of dredged material when possible to maintain the littoral zone, assessment of hydrodynamic processes and changes over time, and evaluation and support of estuary ecosystems.

Anticipating, preparing, responding, and adapting to change all adds up to resilience, and GIS is playing a major role at USACE in supporting resilience on multiple levels through numerous projects: national, regional, and local economic resilience at the mouth of the Columbia River via jetty infrastructure maintenance, large-scale sediment management, marine chart production, environmental resilience for salmon and ecosystems at Trestle Bay, and climate change resilience for coastal infrastructure. It's clear that based on past successes in these areas, GIS will continue to play a crucial role in the engineering portfolio that USACE applies to its multiple roles in ensuring resilience at the mouth of the Columbia River.

REFERENCES

Bonneville Power Administration. 2012. *The Power of the River: The Continuing Legacy of the Bonneville Power Administration in the Pacific Northwest.* Bonneville Power Administration, Portland, OR.

Building climate change resilience in the Pacific Northwest. 2015. Retrieved from https://youtu.be/FWlCA0v74dY.

Center for Economic Development and Research. 2005. "Columbia/Snake River System and Oregon Coastal Cargo Ports Marine Transportation System Study." Prepared by PB Ports and Marine, Portland, OR; BST Associates, Bothell, WA; and Pacific Northwest Waterways Association, Portland, OR.

Columbia River Bar Pilots. www.columbiariverbarpilots.com/columbiariverbarpilots_pilotage.html.

Fisher, Andrew. 2010. *Shadow Tribe: The Making of Columbia River Identity.* Seattle: University of Washington Press.

Haglund, Michael. 2011. "World's Most Dangerous: A History of the Columbia River Bar, Its Pilots and Their Equipment." Columbia River Maritime Museum, Astoria, OR.

Jennings, Todd, Lisa Mighetto, and Jill Schnaiberg. 2003. *Currents of Change: A History of Portland District, U.S. Army Corps of Engineers, 1980–2000.* Portland District, US Army Corps of Engineers.

Lichatowich, Jim. 2001. *Salmon Without Rivers: A History of the Pacific Salmon Crisis.* Washington, DC: Island Press.

Moritz, H. 2016. Interview.

Nienaber Clarke, Jeanne, and Daniel C. McCool. 1996. *Staking Out the Terrain: Power and Performance among Natural Resource Agencies,* Second Edition. Albany, NY: State University of New York Press.

"The Columbia: America's Greatest Power Stream." 1949. Retrieved from https://youtu.be/kXgp8BExpUo.

The White House. 2013, November 1. Executive Order 13653, "Preparing the United States for the Impacts of Climate Change." Washington, DC: Office of the Press Secretary.

US Army Corps of Engineers, Portland District. 2012. "Major Rehabilitation of the Jetty System at the Mouth of the Columbia River." US Army Corps of Engineers, Portland, OR. Retrieved from http://cdm16021.contentdm.oclc.org/cdm/singleitem/collection/p16021coll7/id/3/rec/3.

US Army Corps of Engineers. 2014. "Procedures to Evaluate Sea Level Change: Impacts, Responses and Adaptation." Engineer technical letter 1100-2-1. US Army Corps of Engineers, Washington, DC.

US Army Corps of Engineers, Portland District. 2007. "Summary of Physical and Biological Studies at the Mouth of the Columbia River Sponsored by the U.S. Army Corps of Engineers." US Army Corps of Engineers, Portland, OR.

US Army Corps of Engineers, Portland District. 2015. "Trestle Bay Restoration Project: Final Feasibility Study and Environmental Assessment." US Army Corps of Engineers, Portland, OR. Retrieved from www.nwp.usace.army.mil/portals/24/docs /announcements/fonsi/TrestleBay_Final_FS_EA_July_2015.pdf.

US Army Corps of Engineers, Office of History. 2007. "The U.S. Army Corps of Engineers: A History." Headquarters, US Army Corps of Engineers, Washington, DC.

US Army Corps of Engineers. 2016. "U.S. Army Corps of Engineers: Resilience Initiative Roadmap." Retrieved from http://cdm16021.contentdm.oclc.org/cdm/singleitem /collection/p16021coll6/id/1617/rec/1.

US Census Bureau. 2016. Quick Facts.

US Coast Guard. 2020. Retrieved from Coast Guard Station Cape Disappointment: www.pacificarea.uscg.mil/Portals/8/District_13/lib/doc/factsheet/station_cape_ disappointment.pdf?ver=2017-06-15-151550-343.

Vandy, Greg. 2016. *26 Songs in 30 Days: Woody Guthrie's Columbia River Songs and the Planned Promised Land in the Pacific Northwest.* Seattle: Sasquatch Books.

Willingham, William. 1992. *Army Engineers and the Development of Oregon: A History of the Portland District U.S. Army Corps of Engineers.* Washington, DC: US Government Printing Office.

CHAPTER 5

Urban resilience: Neighborhood spatial complexity and the importance of social connectivity

REGAN M. MAAS

INTRODUCTION

Resilience is the capacity of a place to buffer itself from change and to develop new strategies for survival after disturbances from major natural and human-made disasters. Low-income, socially disorganized neighborhoods are particularly at risk in these situations, as they have fewer resources for assistance after disasters, as well as diminished access to information communication and emotional support through family and friendship networks. However, low-income neighborhoods are not monoliths. Poor neighborhoods can be both economically and environmentally disadvantaged but socially resilient, displaying strong social networks that provide valuable emotional and information support in hard times. Vulnerability and resilience must be measured at the individual level as well as the place level, using the filters of both the site and situation of the neighborhood as well as the social context. Modeling these neighborhoods is a complex process with a rich history. Historically, much of this research has relied on the use of predefined census-based boundaries. However, more modern analyses have capitalized on geospatial science and technology to model these lived spaces in more nuanced ways. Cluster analyses, such as spatial autocorrelation, are a step closer to reality by considering the nature and context of neighboring areas in the construction of neighborhoods. More recently, resident-generated bottom-up measures use information about localized behaviors, activities, and networks to construct more accurate measures of lived space when compared to traditional boundaries. Measuring resilience is not only about resilient cities but also about resilient neighborhoods and resilient individuals. Employing methodologies that consider each of these scales is vitally important. Differentiating neighborhoods based not just on their economic disadvantages but in terms of their social assets will help in finding those areas that are distinctly at risk and those that are more resilient.

SUSTAINABILITY AND RESILIENCE

Over the last several decades, various major environmental disturbances—such as recent natural disasters as well as sociocultural and even economic distur- bances—have thrust into the collective conversation issues about the ability of indi- viduals and communities to respond to these hazards. This has been of particular note, for example, in the Southern California region as municipalities attempt to deal with vulnerabilities inherent in the area, including the risk of wildfire, drought, and major earthquakes and how the diverse urban communities in the region are each uniquely affected. To date, much of the work in this area has centered on ideas of sustainability, which attempt to create and maintain human and natural systems necessary for survival and moderate human needs to fit within the constraints of these systems.

Essentially, the point of sustainability research is to avoid or prevent distur- bances in systems. Although considerable important work continues in sustainabil- ity research, many view this approach as both overly simplistic and unattainable (Berardi 2013; Jickling 1999). Modeling various responses to stressors requires attention to the complexity of closely linked individual, social, and environmental systems and a fundamental understanding that these systems are inherently dynamic (Fiksel 2006; Rees 2010). With this shift in thinking, there is greater interest in the area of resilience research, in which resilience broadly defined is the ability of an individual or community to adapt positively to various stressors (Barr and Devine-Wright 2012; Rees 2010; Folke et al. 2002). It is not the opinion of this author that sustainability and resilience are opposing forces but rather two ways of thinking that are both necessary for devising meaningful and long-term solutions.

DEFINING RESILIENCE

As stated by Rees (2010, 5), "Resilience thinking is a complement to sustainability, not a substitute." Resilience is instead the capacity of a place to buffer itself from change and to learn and develop new strategies for collective survival in the face of disturbances, such as from major natural disasters and even from human-made adversity, as in times of major economic recession (Folke et al. 2002). Overall, there has been a shift in the way we think about these disturbances, from that of vulnerability and deficit to one that emphasizes the varying potentials of communities. This nuance allows for the idea that even those communities that are economically and environmentally at risk may have community-level assets that will mitigate the effects of stressors and allow the communities to be resilient in the face of disturbance.

Various definitions of resilience have developed over time in the literature. These definitions include an understanding of a state of equilibrium that a system is attempting to achieve (May 1973; Holling 1973; Holling 1996; Rose 2004). Some definitions assume that resilient systems will return or bounce back to a state of equilibrium that existed before the introduction of the stressor or disturbance (Holling 1973). Others instead argue that we cannot assume a return to the old equilibrium, but that instead the system will adapt to a new equilibrium or new normal, a point that was previously unknown (Davoudi 2012). All of these arguments, however, agree that there is a threshold or tipping point in the intensity of a disturbance for all systems where equilibrium in any form is unattainable and at which the system inevitably fails (Bennett et al. 2005). Regardless of definition, it is the goal to avoid this tipping point. In the end, resilience measures include the risks individuals and neighborhoods face and the strengths they can use in the face of these risks. These are the ways in which groups of people in different arenas can cope with and adapt to their various sociocultural, economic, and environmental vulnerabilities.

This chapter is specifically aimed at exploring the assets and vulnerabilities of neighborhoods in the vastly complex urban environment. Place matters, and exploring urban environments in terms of both social and physical characteristics is vital to understanding the local variabilities in resilience. Neighborhoods are a meaningful unit of analysis for this discussion, as they are a fundamental reflection of community-level, on-the-ground processes.

THE ROLE OF SPACE IN EQUALITY

The idea that resilience can be meaningfully measured in terms of the assets and vulnerabilities of communities parallels one of public health's central premises regarding the social determinants of health and well-being. Urban neighborhoods that are strong—having both meaningful social connectivity between community members and economic and human capital to access needed resources—are buffered from major stressors, which in turn has a direct effect on well-being (Kawachi and Berkman 2000). Disadvantaged communities, which have little to no effective communication between individuals in the area or strong networks of support and have few physical or economic resources, will be subject to a continuously high stress load, which is strongly linked to chronic health conditions. As expressed by Halfon and Hochstein (2002), health is characterized by an individual's or group's physical ability as well as their capacity to access and use resources, satisfy needs, and thereby cope with environmental influences. It is a complex of social, behavioral, and physical processes. Therefore, resilience of a neighborhood is directly linked to the health of the community, both in terms of physical illness and social cohesion. Measuring vulnerability of a community must consider traditional concerns, such as critical infrastructure, and whether these communities are socially and economically disadvantaged. Such lower-income neighborhoods and their communities must be understood, as they are at particular risk for poor outcomes in the face of disturbance.

Low-income neighborhoods that are also socially disconnected will be at particular risk. For example, during a major natural disaster, such as a large earthquake, low-income areas consistently will have fewer resources to access for assistance, including critical health services. Their limited funds will compound the problem by affording them little economic access to resources that may be available in the greater urban area. If these communities are also socially isolated, in terms of information communication and social support, they face a "double jeopardy" disadvantage.

Income inequality is one specific factor implicated in disrupting the ecological characteristics of social cohesion, or solidarity and connectedness, and levels of trust and reciprocity, or social capital, in a neighborhood (Kawachi and Berkman 2000). Widening the gap between those with money and those without creates inherent differences in environments for individuals in each of these groups, thus affecting health. Income inequality variations between states are highly associated with variations in health outcomes, social factors, and mortality (Kaplan et al. 1996). Income inequality is also highly associated with age-adjusted mortalities. Specifically, mortality declines were smaller in states with greater levels of income inequality (Kaplan et al. 1996).

INCOME INEQUALITY

Income inequality is also implicated in the breakdown of social cohesion in communities and neighborhoods through residential segregation (Kawachi and Kennedy 2002; Kawachi et al. 1997). The spatial concentration of poverty has increased notably since the middle part of the 20th century, concentrating most often in high-density urban communities (Kawachi et al. 1997). As wealthier citizens draw away from the remainder of the citizenry, they often relocate to more affluent areas or bid up the pricing of local housing, forcing lower-income residents out (Kawachi and Kennedy 2002). When the poor are surrounded by others who are also poor, social capital through social networks may be lacking. Community centers and other social organizations may be difficult to establish and maintain because of the lack of both temporal and financial capital in the disadvantaged neighborhood (Wacquant and Wilson 1989).

Income inequality at times encourages social segregation of the disadvantaged, which in turn feeds back to the inequality of the neighborhood, creating intensified social and structural disparities. Frustration, stress, and disruption in these segregated neighborhoods sometimes fuel crime and violence and social disorganization. For example, the spatial concentration of poverty is associated with increased risk of mortality in large urban centers, after controlling for individual income (Kawachi 2002; Waitzman and Smith 1998). Conversely, individuals located in spatially segregated affluent locales were more likely on average to experience lower risks of mortality.

Residential segregation through income inequality negatively affects minority citizens disproportionately. In many metropolitan environments, minorities disproportionately occupy low-income neighborhoods. Residential segregation and racial discrimination in medical care systems and other community resources explain much of the differences found in minority individuals with poor health and general well-being. Communities that experience inequality in income may under-invest in social goods, such as health care and disaster/risk-planning resources (Kennedy, Kawachi, and Prothrow-Stith 1996). Depletion of social capital in the society or the neighborhood often leads to decreased social cohesion, a distinguishing characteristic of socially disorganized neighborhoods (Sampson, Raudenbush, and Earls 1997). The profit motives of many resources, including health and ecological services, also often reinforce the problem. For example, hospitals often relocate out of these disadvantaged areas and health providers often move away from segregated lower-income neighborhoods (Wallace 1990; White-Means 2000).

DISADVANTAGED NEIGHBORHOODS, SPACE, AND RESILIENCE

However, understanding the complexities of these disadvantaged neighborhoods is also vital to appropriately measuring resilience. They are not necessarily monolithic; there is an interaction between economic and social disadvantage across space. We must consider that concentrated poverty and disadvantage do not inevitably create a breakdown in social environments, systems of support and services between individuals and groups, and associated health outcomes, as the traditional public health model would predict. This is especially true for specific minority groups. Not all minority neighborhoods regardless of economic status experience breakdowns in social cohesion, and many retain various social connections, at various scales, which are protective. This is particularly true for Latino communities throughout the United States. The generally established relationship of increased income inequality with decreasing social capital often breaks down in impoverished Latino neighborhoods in US cities (Lara et al. 2005; Markides and Coreil 1986), especially those that are composed largely of foreign-born individuals. Place-based studies that attempt to measure these complexities are needed, and GIS is particularly useful in answering these questions.

Space and place are fundamental parts of the equation when attempting to better understand and model urban resilience (or resilience of any kind). Places are intimately interconnected to sociospatial systems. Greater local community involvement in both public health and resilience interventions is needed to identify their unique resources, needs, and solutions. Geographies at the community level have rarely been considered in attempting to model the neighborhood-level variability of Latino health, for example. Models are frequently designed at the national, regional, state, or county level due to data availability constraints (Markides and Eschbach 2005; Smith and Bradshaw 2006). The varying effects of more spatially local geographies—such as zip codes, census tracts, and especially neighborhoods—are often overlooked (Sastry and Hussey 2003). Investigating such differences at the local scale is challenging and time-consuming due to the resolution of the analysis. However, urban resilience researchers must rise to the challenge, as these areas are a greater reflection of the real lived spaces of individuals from various cultural backgrounds.

LIVED SPACE AND URBAN RESILIENCE

Neighborhoods are fundamentally multidimensional in nature, including both compositional (individual) and contextual (social and physical environmental) aspects. As noted by Cutter (1996), vulnerability is measured at the individual as well as the place levels, where hazard risk mitigation (and resilience) is measured through the filters of both the site and situation of the neighborhood as well as the social context. Modeling neighborhoods in terms of composition alone is limited. Compositional neighborhood effects can be argued as simply aggregations of individual-level effects as opposed to true area-level properties (Anderson, Bulatao, and Cohen 2004). Although compositional characteristics of neighborhoods are integral to understanding and describing social systems within neighborhoods, contextual information is also needed. Area-level factors, including access and availability of resources and support and the ability of individuals to work collectively (collective efficacy), are vital to a thorough understanding of resilience phenomena at the neighborhood level. Additionally, more nuanced considerations of the effects of neighborhood location and context in a given region or city are also needed. According to the Institute of Medicine, "Individuals and families are embedded within social, political, and economic systems that shape behaviors and constrain access to resources necessary to maintain health and well-being" (Institute of Medicine 2001, 241). This embeddedness is true in the natural everyday states of these neighborhoods, so it is incrementally more important in terms of resilience under conditions of disruption.

Understanding the physical environment and planning for hazards is a fundamental element of measuring urban resilience. Measuring the structural assets and deficits in a neighborhood is a complex process vitally important to effective decision-making. These critical infrastructures are those that, if lost, would pose a significant danger to needed resources, including energy and food supplies, and emergency and communications services. Characteristics of a resilient neighborhood are those that consider natural hazard risks in planning decisions for critical infrastructure, that provide secure land tenancy and safe locations for residents, and that systematically require construction that is hazard resilient. As stated by Twigg (2009), the "hardware" approach to resilience takes into account the physical and built environments and their risk assessments but must be accompanied by a software component that includes education in local communities and provides skills training necessary for implementing these systems and policies." GIS-based multicriteria decision support systems are particularly well suited for analyzing these environmentally at-risk communities.

As we now know, neighborhood lived spaces vary considerably and are nuanced even within traditional cultural groups. In these various contexts, individuals will either thrive or fail based strongly on their ability to access and connect to meaningful social networks at the family, community, and regional levels. Various individual factors, ranging from educational attainment to temperament, often mitigate these connections, acting as either barriers or bridges to both social and economic support systems. Therefore, even if a neighborhood has strong overarching social connectivity, individuals may still remain disconnected.

People, in general, by nature attempt to exert control over their situations, regardless of socioeconomic or cultural differences (Bandura 1995). Those individuals who are willingly embedded in or accepting of local expectations of social control that strengthen collective efficacy will therefore be buffered from disruptions themselves by the level of their engagement in the group and their ability to access resources through this connectivity (Stajkovic, Lee, and Nyberg 2009). At the same time, they likely enhance the group itself. Strong social networks will enhance the effectiveness of most phases of the emergency response process during a disaster, including during preparedness phases as well as rescue and recovery (Cutter 2003).

NEIGHBORHOOD SOCIAL NETWORKS AND ETHNICITY

Concentrated disadvantage has long been attributed to decreasing feelings of shared control over aspects of life in the neighborhood, especially regarding problems that are perceived as too large or beyond one's reach. Individual feelings of lack of knowledge about hazards and their impacts specifically, as well as potential actions (individual and collective) that could be protective for the neighborhood, often keep individuals on the sidelines. However, the social control exerted by a group, even if disadvantaged, can be harnessed to bring about change. If interventions are enacted at the neighborhood or collective level, individuals in this system will be more likely to shift behaviors and opinions to that of the collective. This is especially true of ethnic communities, such as the Latino neighborhoods of Los Angeles, California, where strong social connections and identities and behaviors strongly tied to family are pervasive. The networks of communication that are vital for information transfer during disasters are in many ways preformed in many Latino communities where social control is highest and in which strong social connectivity with family and friends exists. Harnessing and enhancing this benefit has particular significance to the current discussion on interventions for resilience.

However, in the end, the true issue at hand is that the benefit of strong social connectivity often subsides for many minority individuals from immigrant populations as they adapt to American society and culture. This dynamic plays out illustratively in the various Hispanic neighborhoods throughout Los Angeles, for example. The intimate intersections between social and economic characteristics of individuals and neighborhoods create a spatial segmentation of the urban landscape. Neighborhoods in the urban core of Los Angeles, which are characterized by high proportions of foreign-born individuals, display strong social capital with extensive family networks (Maas 2016a). Conversely, many first-, second-, and even third-generation individuals who acquire economic capital leave these neighborhoods and move away from the urban core to more affluent suburbs with majority populations of any ethnic background. This spatial shift comes with it a cultural shift with an associated effect on the social systems to which they connect on a regular basis. These individuals are in many ways disconnected, unplugged from a social system that is a protective barrier and a valuable network for information transfer.

Individuals are embedded within social structures. Individual efficacy can be either enhanced or mitigated by the collective beliefs and behaviors of the social systems in which the individual lives. From a sociological perspective, stable social structures and widely held norms are protective and serve to regulate behaviors across myriad individual and collective situations (Durkheim 1897). It is theorized that social networks provide emotional, informational, and material support, regulate behavior, and offer opportunities for engagement, which all benefit individuals and communities (Cassel 1976; Putnam 2001; Cohen 2004; Cohen and McKay 1984; Unger and Powell 1980).

SOCIAL CAPITAL

Discussions of social capital, the cultural capital that focuses on the communal health of neighborhoods—which rely heavily on collective communication, motivation, trust, and behavior—extend the discussion to ideas of social networks and connectedness. In particular, the role of collective efficacy, the combination of social cohesion and shared expectations for beneficial actions on behalf of the community as a critical social process, is relevant to health outcomes as well as a community's ability to work collectively to mitigate the effects of a major disruption (Sampson,

Raudenbush, and Earls 1997; Kawachi and Berkman 2000; Sampson et al. 2002). Robert Putnam's seminal work, "Bowling Alone" (1995), operationalizes social capital as collective ideas of trust and reciprocity as well as measures of the social connections and networks in a community. These measures inherently function at the neighborhood level.

Strong neighboring, or bonding social capital, through a strong local social network, has great importance for the poor and disadvantaged, regardless of race or ethnicity. These connections are especially important in situations where the physical environment is deficient, as in many inner-city situations where quality social and physical health resources are often lacking (Forrest and Kearns 2001). Community environments are often constructed in such a way as to make these interactions difficult. This issue is especially important to urban minorities who are disproportionately burdened by impoverished disordered communities. When neighborhood characteristics of social cohesion and capital begin to break down, many residents wish to move, but their ability to move may be largely constrained by income (Oh 2003). The mitigating effect of social capital may be, however, of less importance to more affluent individuals who can buy in to communities with beneficial environments. Affluent residents may also attract more opportunities to their neighborhoods that are of higher quality.

The characteristics of many urban inner-city environments discourage the development and retention of social capital. Poverty, residential instability, and ethnic heterogeneity are hypothesized to limit the capital with which to support local organizations, impede community attachment through high levels of residential change, and disrupt communication across diverse ethnic groups (Cassel 1976; Browning and Cagney 2003). Segregation of individuals into traditional ethnic low-income communities has been shown to produce neighborhoods with concentrated social dysfunction (Wilkinson and Pickett 2009). These structural factors also affect crime rates despite the ethnic population, signifying that macro-level processes may operate independently of the characteristics of individuals who comprise disadvantaged neighborhoods.

This idea, for example, is pointedly important to the Latino population throughout the United States, but especially in Los Angeles, where large populations of Latinos often live in tightly knit, low-income neighborhoods. Social capital is the means by which they can counteract the normal effects of living in the urban inner city, in terms of health and well-being. Latino populations in lower-income segregated neighborhoods are often composed of large numbers of foreign-born Latinos who retain strong traditional beliefs in the importance of family and community interdependence. Despite the hardships in a disadvantaged Latino neighborhood, many propose that overarching Latino cultural norms retained by recent immigrants, which are highly filial and communal, transcend and push back against the detriments of living in these low-income neighborhoods (Ribble and Keddie 2001; Markides and Coreil 1986; Markides and Eschbach 2005). As expressed by Robert Ramirez in his essay *The Barrio*, "The feeling of the family, a rare and treasured sentiment, pervades and accounts for the inability of the people to leave." (Rosa and Eschholz 2012, 316). In a sense, leaving these neighborhoods could, in fact, be more detrimental, both physically and emotionally, than remaining. This is particularly true for older foreign-born Hispanics/Latinos living in lower-income neighborhoods (Eschbach et al. 2004). It is these central characteristics that must be fully understood when attempting to measure the resilience of a given community, as they extend beyond typical measures that rely heavily on economic ideas of disadvantage. Poor neighborhoods can be both economically and environmentally disadvantaged but socially resilient.

NEIGHBORHOODS: MEASURES OF LIVED SPACE
Census boundaries

Understanding the complexity of resilience in the urban environment requires work that looks not just at large-area regional vulnerabilities and assets but also at small-area contexts that affect individuals directly on a daily basis. In other words, there is a need to identify where people are living and to understand the contexts in which they live in order to determine their overall risks. Regional-, state-, and county-level analyses are not sufficient. A more holistic approach is needed. Additionally, we must understand the benefits of these contexts that are fundamental in determining resilience in the complex urban environment. GIS is particularly suited to analyzing this innately multiscalar complexity. In much of the current research, understanding differences in these contexts often requires operationalizing neighborhoods through delineation of small areal units (Gauvin et al. 2007). Traditional definitions in GIS analyses have used boundaries provided by the US Census, most commonly census tracts, to delineate neighborhood areas. Typically, key variables are aggregated from individual data values to the census boundary to derive contextual data. Inevitably, the use of predefined boundaries, such as census tracts, leads to issues of representativeness and the ability to replicate results. This is the heart of the modifiable areal unit problem that is well-known in geographic analyses (Steinberg and Steinberg 2015).

Census tracts are "compact, recognizable, and homogeneous territorial units with relatively permanent boundaries and an optimum population of about 4,000 people" (US Census Bureau 2010, 7). Specifically, census tracts are designed in an attempt to bound areas that are similar in terms of population and economic characteristics as well as living conditions, as captured by census data. Much of the reason for the large use of the census boundaries is simply consistency in unit of analysis between projects and convenience in terms of connectivity to the large clearinghouse of population data available via the census. It should be noted that the size of US census tracts varies depending on the type of environment. Urban and more densely populated areas tend to have smaller census tracts than rural areas. In many ways, the census tract is the closest systematic predetermined approximation of real neighborhoods available through which to analyze census-based data. However, as noted by Lee et al. (2008), individuals' perceptions of their neighborhoods are often fuzzier than simple tract boundaries. Additionally, many researchers fall into the ecological fallacy trap of making conclusions about individual behavior based solely on analysis at the aggregate level of the tract. More modern attempts at understanding both individual- and neighborhood-level effects have incorporated multilevel or hierarchical modeling techniques to account for this issue while continuing to use census-bounded data as well as new models of the neighborhood, which stretch beyond the simple census boundary. Many of these modeling techniques involve one of many spatial cluster analysis algorithms, which attempt to group areas based on common attributes and spatial characteristics. Mixed methods analyses are also particularly informative in these contexts, by including bottom-up (qualitative) as well as top-down (quantitative) approaches to understanding neighborhood vulnerability and resilience.

Cluster analyses

Various methods attempt to capture more nuanced measures of neighborhoods that look at variability across spaces. Much of this work employs one of many cluster analysis algorithms in a GIS environment, where neighborhoods are defined by identifying spatial relationships between contextual variables across census boundaries. Cluster analysis is a powerful process for grouping data based on similarity (Anselin 1995; Jacquez 2008). It is particularly useful for attempting to delineate neighborhoods. These processes employ the concepts of both spatial heterogeneity and spatial dependence in determining areas whose characteristics are both similar and associated in space, while decreasing similarities between groups. This is in contrast to techniques based on census tracts alone, which rely solely on preexisting boundaries for aggregating population data and are completely blind to neighboring contexts. Delineating neighborhoods not defined simply by census boundaries and that consider the nature and context of neighboring areas when constructing regions is vitally important in a resilience context, as one must naturally sift through the complexity to find both the vulnerabilities and the assets that often lie within and between areas (Anselin 1995; Duque, Anselin, and Rey 2012). Neighborhoods do not exist in a vacuum but are affected by systems (social, economic, and physical) that are not bounded by the arbitrary census boundary.

Spatial autocorrelation is one way of measuring the spatial dependency of similar population characteristics across a given area (Anselin 1995). There are various measures of spatial autocorrelation, including Moran's I, Geary's C, and Getis' G. For example, Moran's I statistic is specifically designed to measure departures from spatial randomness as a means of determining clustering of selected variables (Anselin 1995). Bivariate Moran statistics specifically illustrate the spatial relationships of two census tract-level variables in space. Fundamental to these analyses is a modeling of the underlying spatial connectivity or spatial weights between the geographic units of analysis. For instance, a queen-based spatial weights matrix will measure connectivity based on shared boundaries and vertices. In essence, spatial autocorrelation is high when similar characteristics cluster together in space. Alternatively, spatial autocorrelation is low when dissimilar values cluster.

Fundamentally, most measures of spatial autocorrelation are innately global measures, providing an overarching picture of the spatial association of the given measures across a study region. Using Moran's I once again as an example, the scatterplot can also be used to explore the spatial associations of one or more variables at the local scale, commonly termed *local indicators of spatial autocorrelation* (LISAs) (Anselin 1995).

The four cell matrices of the Moran's scatterplot can be used to split the Moran's I into four categories reflecting local variability in spatial association between the given variable. For example, a univariate Moran's I scatterplot of the Latino population in Los Angeles shows strong global clustering ($I = 0.83$) of this population at the census tract level (figure 5.1). However, by disaggregating and mapping the results of this scatterplot, we can also see where the clustering of tracts that are high in Latino population are located (figure 5.2). In other words, these are census tracts that have high percentages of Latino population surrounded by tracts that are also high in Latino population. However, there are also many Low-Low areas, which are significantly clustered areas that are low in Latino population. Additionally, there are spatial outlier (High-Low and Low-High) areas across the city. Of particular note are those tracts that are high in Latino population but surrounded by non-Latino populations, as they represent spatially isolated populations of Latinos in the city who may experience different neighborhood dynamics from those living in largely clustered Latino areas. Understanding the dynamics of these theoretically culturally isolated neighborhoods would be important in terms of measuring resilience, as isolation may have overarching effects on economic and social connectivity. Adding complexity, bivariate measures of clustering can be used to specifically illustrate the spatial relationships of two census tract-level variables in space. For example, we could visualize both Latino population and income to measure where Latino neighborhoods with higher incomes cluster in space, thus teasing out the spatial complexities of the city, which is important when crafting meaningful neighborhood-level interventions.

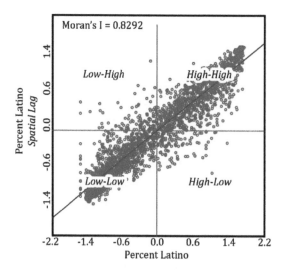

Figure 5.1. A univariate Moran's *I* scatterplot, measuring the spatial autocorrelation (spatial clustering) of the percentage of Hispanics within census tracts throughout Los Angeles in 2010.

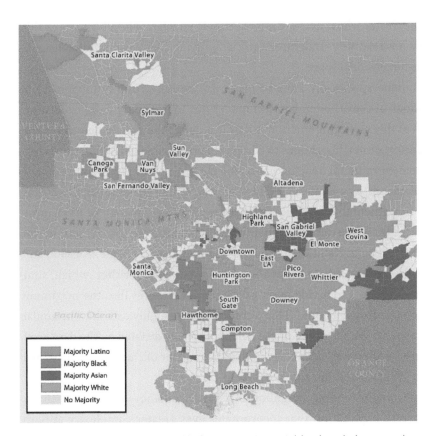

Figure 5.2. Distribution of racial/ethnic majority neighborhoods (greater than 50 percent of the population) in Los Angeles, 2010. Pink areas are majority Latino, green areas are majority Black, purple areas are majority Asian, and blue areas are majority white. Areas with no fill color have no racial/ethnic group that makes up over 50 percent of the population.

However, as we know, neighborhoods are complex constructs that are often defined using a variety of characteristics that stretch beyond just one or even two population characteristics (univariate or bivariate). To more effectively define neighborhoods that reflect these various characteristics, models must be able to detect patterns across large multidimensional datasets (Fincke, Lobo, and Bacao 2008). One long-tested technique for grouping multidimensional characteristics, *k*-means clustering, has been used in many contexts and across many disciplines. *K*-means clustering specifically attempts to group features (often census boundaries in neighborhood research) so that differences between features within a given cluster or neighborhood are minimized. The average of the data characteristics (census variables in many cases) is calculated for each cluster, and each feature is systematically assigned to the cluster whose average it most closely matches. The advantage of this process is that it is relatively straightforward, simplifying complex multidimensional data into two-dimensional space.

This technique specifically aims to maximize within-group homogeneity, a common critique of census boundary techniques due to boundary misspecification. One shortfall of this clustering method is that it requires the user to determine the number of clusters or neighborhoods to compute *a priori*, which is a limitation when attempting to determine the true nature of the clustering in a given complex multidimensional dataset, as are often used in resilience studies. Additionally, *k*-means clustering is fundamentally an aspatial algorithm; however, spatial location information can be added to the calculation to produce spatialized groups. To account for this common criticism in traditional grouping methodologies, Duque, Anselin, and Rey (2012, 2) in recent years developed the max-*p*-regions algorithm, which, similar to *k*-means, attempts to divide "a set of geographic features into the maximum number of homogeneous regions." This technique is also highly data driven, similar again to *k*-means, where the shape of the neighborhood is dictated by the data itself. However, it has an additional advantage in that the number of clusters is determined endogenously (Duque, Anselin, and Rey 2012).

In the end, understanding neighborhoods and resiliency dynamics requires progressive model exploration that incorporates ever greater complexity at both the individual and neighborhoods levels. Therefore, increasing the complexity and meaningfulness of the ways in which neighborhoods, or real on-the-ground lived spaces, are modeled is increasingly important.

Bottom-up models

The neighborhood delineation techniques discussed thus far, regardless of their advancements, still fundamentally rely in practice on assemblages of predefined boundaries in a top-down model, one that defines community from above. This is the fundamental underlying issue defined in the uncertain geographic context problem, as coined by Kwan (2012), where it becomes increasingly difficult, if not impossible, to truly articulate how places affect people if the way in which we define the place or lived space is fuzzy or improperly defined. Resident-generated data models allow the construction of neighborhood lived spaces instead from the bottom up, using information about neighborhood perception and localized behaviors and activities that are more accurate models of lived space when compared with traditional boundaries (Talen 1999; Cutter 2003).

Bottom-up methodologies, which are central to participatory GIS (PGIS) and participatory planning, capture individual differences in lived spaces and can be combined to provide more accurate and, by nature, also more complex representations of neighborhoods (Talen 1999). Understanding individual and collective activity spaces is an important next step in the process of understanding the resilience of individuals and communities, by defining their tangible on-the-ground movements, the places that these movements represent, and their abilities to access needed resources in these complex places. The ability to model these complex systems is an important contribution of GIS and geospatial technology to resilience work, and more is needed. Although area-based work is much of the focus of GIS-based work in public health and hazard vulnerability research, individuals experience space beyond these boundaries in their daily lives, and GIS work must attempt to combine both information at the aggregate as well as the complexity of individual behavior and lived space. Therefore, a combination of both quantitative modeling and qualitative participatory analysis is required to fully understand the places in which individuals live and the assets and exposures found in these places.

LATINO NEIGHBORHOODS IN LOS ANGELES: A CASE STUDY

Neighborhood contexts matter in determining individual and community-level resilience, and contexts are complex, even for populations that are culturally or ethnically similar. Take the case of the predominately Latino neighborhoods of Los Angeles. Often in the academic literature, Latino neighborhoods are treated monolithically in terms of their characterization, especially in terms of a stereotypically "disadvantaged community." However, Latino neighborhoods in Los Angeles vary greatly in character and in space. For these neighborhoods, there is a distinct spatial variability in both economic and social/cultural characteristics. Some Latino neighborhoods are more advantaged than others economically; some have social advantages. Using the various exploratory tools currently available in GIS, we can tease apart these neighborhood contexts in which Latino populations in Los Angeles live.

Latino health paradox

In this case study, the spatial analysis and delineation of Latino neighborhoods was based primarily on the spatial correlation of two variables, immigrant concentration (IC)[1] and concentrated disadvantage (CD), at the census tract level, to map the spatial distributions of both economic and sociocultural characteristics for Latinos in the area (Maas 2016a). Only those census tracts defined as having a majority Latino population (50 percent or more of the tract population is Latino) were considered in the construction of the neighborhood typology. It was proposed that different Latino neighborhood types lead to different social, cultural, and health-related behaviors and outcomes among members in each neighborhood type.

Specifically, as cultural adaptation increases in a neighborhood (as Latinos acculturate to the Los Angeles/American culture), social capital will in turn decrease, leading to increasing stress and poor health outcomes, regardless of the economic fitness of the area. This will in turn have a direct effect on the resilience of both individuals and groups in these neighborhoods. This is central to the well-studied phenomenon found within this population known as the Latino or Hispanic health paradox (Abraido-Lanza et al. 1999; Acevedo-Garcia, Dolores, and Bates 2008). Specifically, disproportionately good health outcomes are routinely observed for Latinos even under economic hardship. However, this paradoxical advantage seemingly wears off the longer these individuals live in the United States and acquire culturally American behaviors, especially in terms of social connectivity. Essentially, individuals living in predominantly Latino foreign-born neighborhoods were experiencing better health outcomes, even though they were living in longstanding economically disadvantaged neighborhoods. The protective effect of their native culture, which places an emphasis on the bonds of family and social connectivity, buffers them from the negative of effects of living in an economically disadvantaged neighborhood.

However, the advantages of living in a wealthier neighborhood cannot be disregarded, especially when planning for major hazard disruptions. It is not simply a factor of measuring social connectivity that determines the ability of a neighborhood to bounce back in times of disruption, although it is decidedly important. It is clear that social connectivity and the bonds provided by social support are vital to a population's ability to resist stressors that are detrimental to health and to collective efficacy (Maas 2016a). However, in terms of the measurement of the hazards that require resilience in these populations, we must look at all of the available assets of the population. It is clear that economic assets and associated high levels of human capital will provide advantages for a place. Individuals in higher income areas are more likely to afford hazard insurance, have knowledge of and implement disaster/risk plans, and have low incidences of psychopathology (Twigg 2009). The effects of both social and economic characteristics will compound. Therefore, economically disadvantaged neighborhoods with dysfunctional social systems will experience cumulative disadvantages. These neighborhoods will understandably be the least resilient in times of disruption, having few tangible (economic) or intangible (social support) assets to harness in order to prepare for, adapt to, and push back against unforeseen stressors.

This cumulative disadvantage is reflected in the recent experiences of the Latino population in the US during the COVID-19 pandemic. In general, Latino individuals make up only 40 percent of the California population, but over 60 percent of the confirmed cases and nearly 50 percent of all deaths from severe acute respiratory syndrome coronavirus 2 (SARS-CoV-2), as of October 2020 (California Department of Public Health 2020). Unfortunately, the existing disparities between advantaged and disadvantaged communities were only exacerbated by the pandemic. Measures to stop the outbreak, including social distancing and community shelter-in-place orders, resulted in disproportionate financial burdens in disadvantaged minority communities where many individuals worked in essential jobs that did not allow work-from-home options and had lower access to medical care and insurance. Compounding the issue, these public health isolation measures also limited the access of these communities to typical sources of social support and social capital, resulting in increased levels of psychological distress and limiting many normal coping mechanisms common within Latino communities (Chow et al. 2020; Fortuna et al. 2020; Hooper, Nápoles, and Pérez-Stable 2020).

Spatial autocorrelation and community: The ethnoburb

Returning to the Los Angeles example, the location of each Latino neighborhood type was defined via the bivariate statistical analysis of Moran's *I* measures of spatial autocorrelation within a GIS environment to find the spatial associations between economic and sociocultural characteristics. The spatial clustering process differentiated the Latino neighborhood landscape of LA, based on both economic and social assets, into four types (figure 5.3): the gateway, the barrio, the enclave, and the ethnoburb[2] (Maas 2016a), casting a light on the great variability of neighborhoods within this group. Traditionally, low-income inner-city Latino neighborhoods can be differentiated by their concentrations of immigrants. Low-income high-immigrant gateway neighborhoods become the first points of contact for many new immigrant individuals, providing culturally safe communities.

Whereas, low-income barrio neighborhoods often contain larger numbers of second- and third-generation Latinos with often weaker connections to cultural and social traditions. Replacing traditional white suburbs as promised lands for residential choice, Latino upper-income neighborhoods also differentiate according to immigrant concentration. Not only are economically advantaged individuals moving to traditional suburbs, but they are also finding spaces in dynamic Latino suburbs. However, these new suburbs are not singular in form. New Latino suburbs differentiate in their concentrations of immigrants. Latino immigrant suburbs or enclaves are characterized by stronger economies than their gateway counterparts, with the development of viable ethnic business districts in these neighborhoods. Low-immigrant Latino ethnoburb suburbs, by contrast, display even stronger economies both within the neighborhood and in their connections to the host system at large, similar to Chinese ethnoburbs in the San Gabriel Valley in Los Angeles originally defined in Li (2009).

Spatialized segregation

When attempting to understand the vulnerability and resilience of a neighborhood, we must dig deeper to see the differences within groups. In Los Angeles, there is considerable spatial segmentation of Latino neighborhoods across the area (figure 5.4). There are large populations of Latinos in both the highly immigrant disadvantaged gateway neighborhoods in the central city as well as in the more economically advantaged nonimmigrant ethnoburb. There is considerable evidence to show that these two neighborhood types have grown considerably since the year 2000 (Maas 2016b). All these neighborhoods are majority Latino, but they are very different places. The gateway neighborhoods of the inner city are predominately low-income; however, research (Maas 2016a) has shown that the highly immigrant homogeneous cultural environment of these neighborhoods is protective in terms of health and well-being by having strong social connectivity and systems of support. Conversely, the ethnoburb is economically quite strong, but social networks are strained, as traditional cultural identities begin to fade away with increasing acculturation. Unfortunately, the barrio experiences the compounding effects of both economic and sociocultural disadvantage, making individuals disconnected from financial and physical resources and services but also disconnected from networks of information and support. Using this typology then, the enclave is most advantaged in the spectrum of Latino neighborhoods, having theoretically both stronger social connectivity as well as a stronger economy, making it increasingly more resilient. Unfortunately, few of the Latino neighborhoods in Los Angeles are of this type.

However, it must be noted that this does not mean these areas are more advantaged than an affluent predominantly white community. What this does mean is that some Latino neighborhoods are more buffered than others and potentially more resilient. This is most likely true of many racial or ethnic groups who are often treated monolithically. Place matters and modeling the complexity of place in resilience studies is more important than ever. The ability of GIS to capture and sift through this complexity is needed in future work.

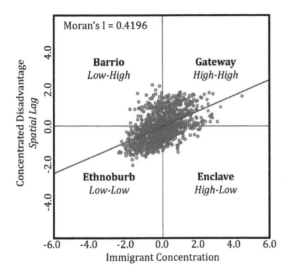

Figure 5.3. A bivariate Moran's *I* scatterplot neighborhood typology using local spatial association between immigrant concentration and concentrated disadvantage across Latino neighborhoods in Los Angeles.

An even deeper examination of the spatial variability between social and economic characteristics using the max-p-regions grouping methodology reveals a refinement of the neighborhood delineations, providing an even closer approximation of the on-the-ground dynamics (figure 5.5). The four-pronged typology has now been disaggregated into 238 unique areas based on the full spectrum of values in both immigrant concentration and concentrated disadvantaged, as opposed to simply highs and lows. To illustrate the point, the relatively uniform city of West Covina (a California ethnoburb in figure 5.4) has now been grouped into 18 new

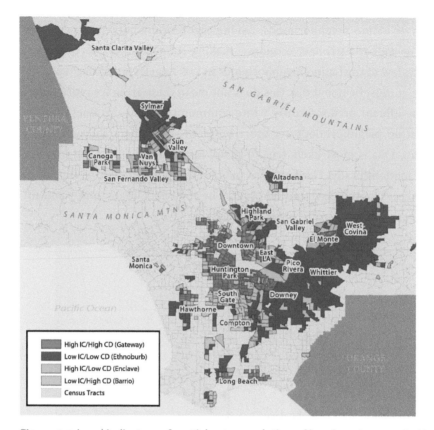

Figure 5.4. Local indicators of spatial autocorrelation of immigrant concentration and concentrated disadvantage across majority Latino neighborhoods in the Los Angeles area. Red areas represent both high IC and CD (gateway communities), pink areas represent low IC and high CD (barrio), dark-blue areas represent both low IC and CD (ethnoburb), and light-blue areas represent high IC and low CD (enclaves).

neighborhoods, all of which will be characteristically different. These illuminations of complexity make the process of modeling vulnerability messier but are no less needed.

However, to truly understand the nature of the contexts and places that Latino individuals in LA navigate, we need a resident-generated data approach. Studies that examine the spatial structures of these individuals' daily activity spaces via space-time models are needed (Kwan 1998). As noted by Kwan (1998), although traditionally performed using pen-and-paper travel diary records, new investigations are using GPS-enabled smartphone applications to collect location-based data on the geographic neighborhood boundaries of activity spaces for individuals, as well as, motivations and behaviors related to node preference in the activity space, and the

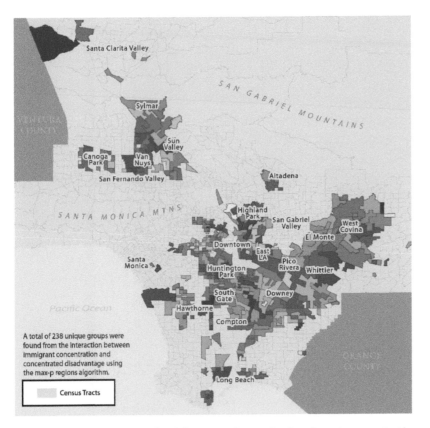

Figure 5.5. Max-*p*-regions algorithm groupings using immigrant concentration and concentrated disadvantage across majority Latino neighborhoods in the Los Angeles area. A total of 238 unique groups were found from the interaction between immigrant concentration and concentrated disadvantage using the max-*p*-regions algorithm.

structures of social networks in these spaces (figure 5.6). These activity paths can then be used to find spatial relationships at the individual level or overlaid and generalized across a group in a GIS to find patterns. These analyses can help determine the spatial relationships between these new neighborhood constructions and the opportunities and vulnerabilities in these spaces.

This research is being used to construct a more nuanced model of Latino neighborhood typology that considers variability both spatially and temporally. Definitions of neighborhoods less constrained by preset boundaries and using fuzzier definitions that consider the movement of individuals through space and time provide more meaningful ways to model Latino environments.

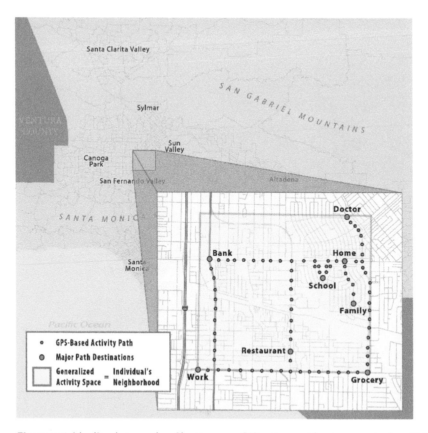

Figure 5.6. Idealized example of bottom-up GIS using resident-generated participatory GPS smartphone activity space data capture. Large circles represent major destinations in the community and small circles represent GPS-based activity paths.

Best practice for identifying resilient communities

Some groups are more resilient than others, but these differences are not controlled by race or ethnicity or even necessarily socioeconomic conditions. In some communities, especially those composed largely of recent immigrants, cultural traditions (and even pressures) that emphasize family and social connectivity are written into the fabric of the neighborhood. These traditional values include cohesiveness and positive attachment, support and feelings of obligation to the overall group, and community involvement, as well as complex networks for the communication of information about important resources.

Resilience is not controlled by the social constructions of race or ethnicity. This is not to say that race and ethnicity do not have a role, as these characteristics still remain strongly tied to economic advantages (or disadvantages) that are important resources when attempting to understand a neighborhood's capacity to rebound in the face of major disturbances. However, when we dig deeper into the sociocultural and spatial complexities of these groups across the urban environment, there are notable differences. For example, some Latino neighborhoods are highly socially connected, while others are not. As spatial analyses of neighborhoods show, the contexts of neighborhoods are clearly implicated in assessing the resilience of a given community. Living in neighborhoods where norms reinforce the importance of family and social connectivity will be inherently protective. Therefore, quantifying and identifying these specific protective neighborhoods is vital for analyses that attempt to model vulnerability versus resilience at this scale.

Examining urban resilience in the future will require examining the complex multidimensional characteristics (economic, social, and so on) of the urban environment. Resilience must also be measured at multiple scales, including the regional and the local. We must also be cognizant of the complexity of how we delineate or bound communities and their lived spaces. Current methods that rely on preexisting boundaries are often weak and cannot adequately reflect the contexts that directly affect their everyday lives—thus the need for a more local, place-based definition of community arises. Geographic information systems are central to this multiscalar, multidimensional understanding of vulnerability and resilience, as it inherently allows for the integration and analysis of complex spatial datasets.

FINAL THOUGHTS

It is not just about resilient cities but also about resilient neighborhoods and resilient individuals. Yes, we must have city- and countywide efforts that are aimed at securing infrastructure, planning and management of emergency services, and reliable communication systems, but we must also evaluate resilience at the neighborhood level in terms of both the obvious built environment but also in terms of the social environment. Positive social environments are assets for the dissemination of information in times of crisis, for providing social support and enhancing the collective action of the group, and for pushing back against countless stressors experienced during times of disruption. Differentiating neighborhoods based not just on their economic disadvantages—which would systematically classify Latino neighborhoods, for example, as vulnerable—but also in terms of their social assets will help in finding those areas that are truly at risk and those that are resilient.

REFERENCES

Abraido-Lanza, Ana F., Bruce P. Dohrenwend, Daisy S. Ng-Mak, and J. Blake Turner. 2009. "The Latino Mortality Paradox: A Test of the Salmon Bias and Healthy Migrant Hypotheses." *American Journal of Public Health* 89 (10): 1543–48.

Acevedo-Garcia, Dolores, and Lisa M. Bates. 2008. "Latino Health Paradoxes: Empirical Evidence, Explanations, Future Research, and Implications." In *Latinas/os In The United States: Changing The Face Of America*, 101–113. New York City: Springer US.

Anselin, Luc. 1995. "Local Indicators of Spatial Association – LISA." *Geographical Analysis* 27: 93–115.

Bandura, Albert. 1995. "Exercise of Personal and Collective Efficacy in Changing Societies." In *Self-efficacy in Changing Societies*, edited by Albert Bandura, 1–38. Cambridge University Press.

Barr, Stewart, and Patrick Devine-Wright. 2012. "Resilient Communities: Sustainabilities in Transition." *Local Environment* 17 (5): 525–32.

Bennett, E. M., G. S. Cumming, and G. D. Peterson. 2005. "A Systems Model Approach to Determining Resilience Surrogates for Case Studies." *Ecosystems* 8 (8): 945–57.

Berardi, Umberto. 2013. "Clarifying the New Interpretations of the Concept of Sustainable Building." *Sustainable Cities and Society* 8: 72–78.

Browning, Christopher R., and Kathleen A. Cagney. 2003. "Moving Beyond Poverty: Neighborhood Structure, Social Processes, and Health." *Journal of Health and Social Behavior* 44 (4): 552–71

California Department of Public Health. *COVID-19 Race and Ethnicity Identity.* https://www.cdph.ca.gov/Programs/CID/DCDC/Pages/COVID-19/Race-Ethnicity.aspx. Accessed Oct. 7, 2020.

Cassel, John. 1976. "The Contribution of the Social Environment to Host Resistance." *American Journal of Epidemiology* 104 (2): 107–23.

Chow, Daniel S., Jennifer Soun, Justin Gavis-Bloom, Brent Weinberg, Peter Chang, Simukayi Mutasa, Edwin Monuki, Jung In Park, Xiaohui Xie, Daniela Bota, Jie Wu, Leslie Thompson, Alpesh Amin, Saahir Khan, and Bernadette Boden-Albala. 2020. "The Disproportionate Rise in COVID-19 Cases among Hispanic/Latinx in Disadvantaged Communities of Orange County, California: A Socioeconomic Case-Series." *BMJ Yale* (preprint).

Cohen, Sheldon. 2004. "Social Relationships and Health." *American Psychologist* 59 (8): 676–84.

Cohen, Sheldon, and Garth McKay. 1984. "Social Support, Stress and the Buffering Hypothesis: A Theoretical Analysis." *Handbook of Psychology and Health* 4: 253–67.

Cutter, Susan L. 1996. "Societal Responses to Environmental Hazards." *International Social Science Journal* 48 (150): 525–36.

Cutter, Susan L. 2003. "GIScience, Disasters and Emergency Management." *Transactions in GIS* 7 (4): 439–44.

Davoudi, S. 2012. "Resilience: A Bridging Concept or a Dead End? Applying the Resilience Perspective to Planning: Critical Thoughts from Theory and Practice," edited by Simin Davoudi and Libby Porter. *Planning Theory & Practice* 13 (2): 299–333.

Duque, Juan C., Luc Anselin, and Sergio J. Rey. 2012. "The Max-*p*-Regions Problem." *Journal of Regional Science* 52 (3): 397–419.

Durkheim, Emile. 1897. *Le Suicide: Étude de Sociologie*, edited by F. Alcan.

Eschbach, Karl, Glenn V. Ostir, Kushang V. Patel, Kyriakos S. Markides, and James S. Goodwin. 2004. "Neighborhood Context and Mortality Among Older Mexican Americans: Is There a Barrio Advantage?" *American Journal of Public Health* 94 (10): 1807–12.

Fiksel, Joseph. 2006. "Sustainability and Resilience: Toward a Systems Approach." *Sustainability: Science, Practice & Policy* 2 (2): 14–21.

Fincke, Tonio, Victor Lobo, and Fernando Bacao. 2008. "Visualizing Self-Organizing Maps with GIS." GI-Days 2008: Münster.

Folke, Carl, Steve Carpenter, Thomas Elmqvist, Lance Gunderson, Crawford S. Holling, and Brian Walker. 2002. "Resilience and Sustainable Development: Building Adaptive Capacity in a World of Transformations." *AMBIO: A Journal of the Human Environment* 31 (5): 437–40.

Forrest, Ray, and Ade Kearns. 2001. "Social Cohesion, Social Capital and the Neighborhood." *Urban Studies* 38 (12): 2125–43.

Fortuna, Lisa R., Marina Tolou-Shams, Barbara Robles-Ramamurthy, and Michelle Porche. 2020. "Inequity and the Disproportionate Impact of COVID-19 on Communities of Color in the United States: The Need for a Trauma-Informed Social Justice Response." *Psychological Trauma: Theory, Research, Practice, and Policy* 12 (5): 443–45.

Gauvin, Lise, Eric Robitaille, Mylène Riva, Lindsay McLaren, Clement Dassa, and Louise Potvin. 2007. "Conceptualizing and Operationalizing Neighborhoods: The Conundrum of Identifying Territorial Units." *Canadian Journal of Public Health / Revue Canadienne de Santé Publique* 98 (S1): S18–S26.

Halfon, Neal, and Miles Hochstein. 2002. "Life Course Health Development: An Integrated Framework for Developing Health, Policy, and Research." *Milbank Quarterly* 80 (3): 433–79.

Holling, Crawford Stanley. 1973. "Resilience and Stability of Ecological Systems." *Annual Review of Ecology and Systematics* 4: 1–23.

Holling, Crawford Stanley. 1996. "Engineering Resilience versus Ecological Resilience." In *Engineering Within Ecological Constraints*, edited by Peter Schulze, 31–44. Washington, DC: National Academy of Engineering, National Academies Press.

Hooper, Monica W., Anna María Nápoles, and Eliseo J. Pérez-Stable. 2020. "COVID-19 and Racial/Ethnic Disparities." *JAMA*, 323 (24): 2466–67.

Institute of Medicine. 2001. "Health and Behavior: The Interplay of Biological, Behavioral, and Societal Influences." In *Committee on Health and Behavior, Research, Practice, and Policy, Board on Neuroscience and Behavioral Health.* Washington, DC: National Academies Press.

Jacquez, Geoffrey M. 2008. "Spatial Cluster Analysis." In *The Handbook of Geographic Information Science,* edited by John P. Wilson and A. Stewart Fotheringham, 395–416.

Jickling, Bob. 1999. "Beyond Sustainability: Should We Expect More From Education?" *Southern African Journal of Environmental Education* 19: 60–7.

Kaplan, George A., Elsie R. Pamuk, John W. Lynch, Richard D. Cohen, and Jennifer L. Balfour. 1996. "Inequality in Income and Mortality in the United States: Analysis of Mortality and Potential Pathways." *British Medical Journal* 312 (7037): 999–1003.

Kawachi, Ichiro. 2002. "Income Inequality and Economic Residential Segregation." *Journal of Epidemiology and Community Health* 56: 165–66.

Kawachi, Ichiro, and Lisa F. Berkman. 2000. "Social Cohesion, Social Capital, and Health." In *Social Epidemiology,* edited by Lisa F. Berkman and Ichiro Kawachi, 174–190.

Kawachi, Ichiro, and Bruce P. Kennedy. 2002. *The Health of Nations: Why Inequality Is Harmful to Your Health.* New York, NY: The New Press.

Kawachi, Ichiro, Bruce P. Kennedy, Kimberly Lochner, and Deborah Prothrow-Stith. 1997. "Social Capital, Income Inequality, and Mortality." *American Journal of Public Health* 87 (9): 1491–98.

Kennedy, Bruce P., Ichiro Kawachi, and Deborah Prothrow-Stith. 1996. "Income Distribution and Mortality: Cross-Sectional Ecological Study of the Robin Hood Index in the United States." *British Medical Journal* 312 (7037): 1004–1007.

Kwan, Mei-Po. 1998. "Space-Time and Integral Measures of Individual Accessibility: A Comparative Analysis Using a Point-Based Framework." *Geographical Analysis* 30 (3): 191–216.

Lara, Marielena, Cristina Gamboa, M. Iya Kahramanian, Leo S. Morales, and David E. Hayes Bautista. 2005. "Acculturation and Latino Health in the United States: A Review of the Literature and Its Sociopolitical Context." *Annual Review of Public Health* 26: 367–97.

Lee, Barrett A., Sean F. Reardon, Glenn Firebaugh, Chad R. Farrell, Stephen A. Matthews, and David O'Sullivan. 2008. "Beyond the Census Tract: Patterns and Determinants of Racial Segregation at Multiple Geographic Scales." *American Sociological Review* 73 (5): 766–91.

Li, Wei. 2009. *Ethnoburb: The New Ethnic Community in Urban America.* Honolulu, HI: University of Hawaii Press.

Maas, R. 2016a. "Breaking Bonds: The Hispanic Health Paradox in Los Angeles." *Yearbook of the Association of Pacific Coast Geographers* 78: 120–47.

Maas, R. 2016b. "The Neighborhood Dynamics of Hispanic/Latino Communities in Los Angeles: 2000 to 2010." Presentation at the annual meeting of the American Association of Geographers, San Francisco, CA, April 29–30, 2016.

Markides, Kyriakos S., and Jeannine Coreil. 1986. "The Health of Hispanics in the Southwestern United States: An Epidemiologic Paradox." *Public Health Reports* 101 (3): 253–65.

Markides, Kyriakos S., and Karl Eschbach. 2005. "Aging, Migration, and Mortality: Current Status of Research on the Hispanic Paradox." *The Journals of Gerontology Series B: Psychological Sciences and Social Sciences* 60 (Special Issue 2): S68–S75.

May, Robert M. 1973. "Qualitative Stability in Model Ecosystems." *Ecology* 54 (3): 638–41.

Morenoff, Jeffrey D., and John W. Lynch. 2004. "What Makes a Place Healthy? Neighborhood Influences on Racial/Ethnic Disparities in Health over the Life Course." In *Critical Perspectives on Racial and Ethnic Differences in Health in Late Life*, edited by N. B. Anderson, R. A. Bulatao, and B. Cohen. National Research Council (US) Panel on Race, Ethnicity, and Health in Later Life. Washington, DC: National Academies Press.

Oh, Joong-Hwan. 2003. "Social Bonds and the Migration Intentions of Elderly Urban Residents: The Mediating Effect of Residential Satisfaction." *Population Research and Policy Review* 22 (2): 127–46.

Putnam, Robert D. 1995. "Bowling Alone: America's Declining Social Capital." *Journal of Democracy* 6 (1): 65–78.

Putnam, Robert. 2001. "Social Capital: Measurement and Consequences." *Canadian Journal of Policy Research* 2 (1): 41–51.

Rees, William E. 2010. "Thinking 'Resilience.'" In *The Post Carbon Reader: Managing the 21st Century's Sustainability Crises*. Santa Rosa, CA: Watershed Media, Post Carbon Institute.

Ribble, John C., Luisa Franzini, and Arlene M. Keddie. 2001. "Understanding the Hispanic Health Paradox." *Ethnicity & Disease* 11: 496–518.

Rosa, Alfred, and Paul Eschholz. 2012. *Models for Writers: Short Essays for Composition, 11th Edition*, 316–22. Boston, MA and New York, NY: Bedford/St. Martin's.

Rose, Adam. 2004. "Defining and Measuring Economic Resilience to Disasters." *Disaster Prevention and Management: An International Journal* 13 (4): 307–14.

Sampson, Robert J., Stephen W. Raudenbush, and Felton Earls. 1997. "Neighborhoods and Violent Crime: A Multilevel Study of Collective Efficacy." *Science* 277 (5328): 918–24.

Sastry, Narayan, and Jon M. Hussey. 2003. "An Investigation of Racial and Ethnic Disparities in Birth Weight in Chicago Neighborhoods." *Demography* 40 (4): 701–25.

Smith, David P., and Benjamin S. Bradshaw. 2006, "Rethinking the Hispanic Paradox: Death Rates and Life Expectancy for US Non-Hispanic White and Hispanic Populations." *American Journal of Public Health* 96 (9): 1686–92.

Stajkovic, Alexander D., Dongseop Lee, and Anthony J. Nyberg. 2009. "Collective Efficacy, Group Potency, and Group Performance: Meta-Analyses of their Relationships, and Test of a Mediation Model." *Journal of Applied Psychology* 94 (3): 814–28.

Talen, Emily. 1999. "Constructing Neighborhoods from the Bottom Up: The Case for Resident-Generated GIS." *Environment and Planning B: Planning and Design* 26 (4): 533–54.

Twigg, John. 2009. "Characteristics of a Disaster-Resilient Community: A Guidance Note" (version 2 November). For the DFID Disaster Risk Reduction Interagency Coordination Group.

Unger, Donald G., and Douglas R. Powell. 1980. "Supporting Families Under Stress: The Role of Social Networks." *Family Relations* 29 (4): 566–74.

US Census Bureau. United States Census 2010: Participant Statistical Areas Program Guidelines. Washington, DC: US Census Bureau.

Wacquant, Loïc J. D., and William Julius Wilson. 1989. "The Cost of Racial and Class Exclusion in the Inner City." *The Annals of the American Academy of Political and Social Science* 501 (1): 8–25.

Waitzman, Norman J., and Ken R. Smith. 1998. "Separate but Lethal: The Effects of Economic Segregation on Mortality in Metropolitan America." *Milbank Quarterly* 76 (3): 341–73.

Wallace, Steven P. 1990. "The Political Economy of Health Care for Elderly Blacks." *International Journal of Health Services* 20 (4): 665–80.

White-Means, Shelley I. 2000. "Racial Patterns in Disabled Elderly Persons' Use of Medical Services." *The Journals of Gerontology Series B: Psychological Sciences and Social Sciences* 55 (2): S76–S89.

Wilkinson, Richard G., and Kate E. Pickett. 2009. "Income Inequality and Social Dysfunction." *Annual Review of Sociology* 35: 493–511.

NOTES

1   Immigrant concentration (IC) and concentrated disadvantage (CD) were derived from factor analyses designed using data from the 2010 US Census. IC estimates the number of immigrants in the census tract by considering the percentage of those who were Latino, were born outside the United States and immigrated since 2000, who spoke Spanish, and were not US citizens. CD estimates by considering households that lived in poverty, were headed by a female alone with children, lived on public assistance, had incomes less than $24,000, and were majority composed of children under 18.

2   Li, 2009.

CHAPTER 6

Community resilience, contested spaces, and Indigenous geographies

DEAN OLSON, ALLISON FISCHER-OLSON, BRENDA NICOLAS, WENDY TEETER, MAYLEI BLACKWELL, AND MISHUANA GOEMAN

INTRODUCTION

A map of Los Angeles does not tell the story of its original people. In a large city such as Los Angeles, this is a story that is often invisible to policy makers and even the city's notion of itself. Los Angeles has the largest Indigenous population of any city in the US,[1] though these communities are often invisible to each other. The Mapping Indigenous Los Angeles (MILA) project exposes a layered geography of Indigenous Los Angeles that includes Gabrielino-Tongva and Fernandeño/Tataviam who struggle for recognition and protection of their original spaces, American Indians who were removed from their lands and displaced through governmental policies of settler colonialism,[2] and Indigenous diasporas[3] from Latin America and Oceania that have been displaced by militarism, neoliberal economic policies, and overlapping colonial histories.[4] The unique aspect of this project is that it uses participatory geographic information systems (GIS) and collaborative engagement to tell the stories of Indigenous groups that live in the Los Angeles area.

From the mainstream geographic perspective, Los Angeles is a complex and unique city with incredible diversity, thanks to its distinctive cultural and economic position. Widely known as the entertainment capital of the world, Los Angeles is the second-largest city in the United States and the seat of the nation's largest county in terms of population.[5] Economic activity accelerated dramatically in the postwar era in large part due to the region's burgeoning aerospace industry and busy shipping port. The region's seemingly limitless supply of flat land provided ample space for suburban development, which had been spurred on by various federal policies immediately following World War II (Housing Act of 1949, Federal Aid Highway Act of 1956, VA home loans).[6] The confluence of growth and money during that period transformed Los Angeles into the sprawling metropolis it is today. Unlike older cities, which tend to emphasize a single center that fades outward, Los Angeles is characterized by multiple centers with distinctive personalities and characteristics. Often, Los Angeles neighborhoods are characterized by the cultural-ethnic makeup of their inhabitants. Over the years, places in Los Angeles were named using conventions such as Koreatown, Little Armenia, and Little Ethiopia. In fact, place is constantly being made for new and old immigrant communities in the city. On the other hand, there is little acknowledgement of the city's Indigenous roots. The MILA project provides that voice.[7] It was conceived by UCLA scholars of Indigenous Studies, Maylei Blackwell (Cherokee), Mishuana Goeman (Tonawanda Band of Seneca), Keith Camacho (Chamorro), and Wendy Teeter. This project challenges two common ways of representing Los Angeles: the textbook narrative and the usual Web Mercator projection.

MAPPING INDIGENOUS LOS ANGELES

MILA instead looks at the landscape of Los Angeles for its many layers. Following Indigenous protocol—ways of consulting with Indigenous individuals and communities to maintain positive and respectful relationships—in this case begins by acknowledging those on whose land the project is based. While many would argue that there is not one Los Angeles but multiple, what is less known is that there are many Indigenous Los Angeles communities whose histories are layered into the fabric of the city. The MILA project attempts to show how the original peoples of the Los Angeles basin (and islands) relate specifically to this land and how subsequent relocations and migrations of Indigenous peoples have reworked space, place, and meaning within a shared landscape. While land itself is material, and thus an important aspect in navigation through space, landscape involves various lenses through which people view the land and ascribe location-specific meaning. People ascribe meaning to the places in which they live, and the environment in turn informs an element of cultural identity. In Indigenous landscapes, culturally based views are informed by epistemologies that include the land, humans, and nonhumans, who also exercise agency. These geographies are an important part of human identity, and land cannot be simply abstracted from its inhabitants. People and nonhuman agents interact with the land and with each other extensively for recreation, sustenance, shelter, and other necessities, and this interactive process creates a multidirectional informative system. Through MILA, we aim to uncover the multiple layers of Indigenous Los Angeles landscapes through a digital storytelling and oral history project with community leaders and elders from Indigenous communities throughout the city. Working for, with, and in Indigenous communities, the project creates a platform available to a broad public audience for communities to use in ongoing endeavors of self-representation and continual assertions of place making.

A layered Indigenous Los Angeles

To provide adequate context to the issues addressed by the project, this chapter begins with a brief overview of historical and contemporary contexts, as necessary, of various layers of the Indigenous Los Angeles landscape that MILA addresses in the project specifically. Next, the chapter discusses critical cartography/geography and the role of GIS in presenting invisible histories and communities. After a brief introduction to working with and in communities, three case studies are presented: "Mapping Tongva LA," "Making a Collaborative LAID Map," and "Urban Native LA." These case studies will outline issues, successes, and challenges in working with three different Indigenous Angeleno communities to create ArcGIS® StoryMaps℠ *stories* (sometimes known as *story maps*). These maps are self-representations of specific communities and thus different capacities were used in the StoryMaps environment. The chapter wraps up in a final discussion about the project's overall experience at this stage using the StoryMaps app, including pros and cons, and next steps the MILA project hopes to undertake.

Mapping Indigenous Los Angeles begins by uncovering stories of Indigenous Los Angeles layer by layer, starting with those who first lived in the region. The first peoples of the Los Angeles basin and San Fernando Valley (Gabrielino-Tongva and Fernandeño/Tataviam) maintain distinct geographies of place, community, and identity. It is a consistent struggle to have their identities and relationships to this place recognized by the wider public. For example, this struggle is highlighted in Puvungna, the birthplace of Chinigchinich, located on the campus of California State University, Long Beach (CSULB). In 1992, CSULB attempted to convert 22 acres into a strip mall. The then-76-year-old Lillian Robles, a respected Acjachemen elder, along with Jim Alvitre (Tongva) and many others, occupied the land for two weeks, facing off with police, university officials, developers, and bulldozers.[8] In response to situations such as this, Acjachemen/Juaneño and Gabrielino-Tongva organized an annual Ancestor Walk. This event has become a pilgrimage to call attention to their endangered and lost sacred sites across Los Angeles and Orange Counties such as Panhe, Putiidhem, Kenyaanga, Bolsa Chica, Motuuecheynga, and Puvungna. It also highlights the present and remembered geography of these first peoples.

The project seeks to recover a second layer of memory of American Indians who came to the city as part of the "relocation days,"[9] focusing on historical narratives of Native Los Angeles and the places where these stories originate. Los Angeles became the largest urban Native American population due to a period known as relocation and termination, which took place in the 1950s under Dillon S. Myer, then the Commissioner of the Bureau of Indian Affairs. The United States government implemented a specific policy of termination of reservations and "detribalization" to relocate reservation-based populations to urban centers to participate in industrial wage-earning jobs. Faced with relocation and termination policies, poor economic conditions, and staggering unemployment on reservations in the 1940s–1960s, American Indians moved en masse off reservations. This was only the first wave, as conditions have not changed and people continue to move to where they can find employment, receive an education, and join family. In some cases, for more than three generations, families have moved back and forth between Los Angeles and reservations, forming complex networks throughout the city and space in between.[10] Whole streets started to develop that were composed of relatives, which makes LA a unique urban Indian community to explore. Unlike ethnic enclaves or other community settlement patterns, urban Indians have not had one geographic locale but are often dispersed throughout the city. This part of the research is firmly situated in American Indian studies, yet it has ties to the way in which Pacific Islanders and Mexican and Latin American Indigenous peoples also become displaced and then make place in Los Angeles. Making place is a process by which diasporic communities re-create and maintain a sense of identity and belonging in a new location away from home, whether it be relationships with other people, animals, deities, the elements, or the land. The idea of making place is multifaceted, taking spatial forms such as delineation of safe spaces that serve as community focal points for culturally specific wellness or educational services, and programming, among other gatherings—all important in a city such as Los Angeles, where Indigenous communities are widely distributed or may be invisible to each other.

Los Angeles–an Indigenous hub

Given the intensified flow of peoples, capital, and cultures due to neoliberal global economic restructuring, Los Angeles is a transnational hub[11] for many people and migrant streams, specifically for an increasingly Indigenous diaspora from Mexico and Guatemala. While Los Angeles has a rich Latino immigrant history, public services fail this ever-growing population. Immigrants are often non-Spanish speaking and monolingual speakers of their own Indigenous language. This was dramatically illustrated by the Los Angeles Police Department (LAPD) shooting death of Manuel Jaminez Xum, a Mayan (K'iche') day laborer, in the Westlake neighborhood of Los Angeles in 2010 when it was not clear that he only spoke K'iche' and thus did not understand the police commands as a non-English/non-Spanish speaker. The estimated 120,000 Oaxacan Indigenous immigrants settling and working in Los Angeles challenge the larger US society to see "Mexican" immigration as a multiracial process.[12] Along with the growing diaspora of Mayans (largely Kanjobal and K'iche') from Guatemala, their presence shakes loose some of the perceived notions of Latinidad and indigeneity. This research, then, interrogates ideas of race and racialization in relation to indigeneity and migration from the point of view of Indigenous community activists throughout LA.[13] Further, our collaboration examines dynamic, converging geographies of Native Los Angeles, such as ceremonial sites used by the Mayan Daykeepers Association, that create ongoing forms and newly articulated geographies.

The fourth part of this project focuses on Pacific Islanders in Los Angeles. California boasts the largest number of Chamorros, Hawaiians, Marshallese, Samoans, and Tongans outside of Oceania, at 286,145 as of 2010.[14] According to the 2010 census, 105,348 Pacific Islanders live in the Los Angeles Combined Statistical Area, with more than one-third residing in Los Angeles County alone.[15] Through the MILA project, we explore Pacific Islanders' understanding of and engagement with the land as an urban environment, as well as how Islanders construe and construct their respective notions of indigeneity in a new place. Pacific Islanders have long interacted with the peoples of the Americas, from economic exchanges and political conflicts with Spanish missionaries in the 17th century to cultural and martial appropriations of American soldier masculinities in the 20th century. While the historical records and scholarly assessments of these processes make for important discussions, seldom have these debates analyzed Pacific Islander migration to and settlement in the Americas. The project addresses these questions by turning to the expertise of Pacific Islander elders and community organizers throughout the greater Los Angeles area.

Telling a multilayered Indigenous story

The MILA project spans multiracial contexts of Los Angeles and reveals cultures often hidden or obscured within nationalized cultures that are recognized by most city planners and scholars of the city. While many Indigenous Angelenos might take issue with being portrayed as emerging, as this characterization erases long histories in the city (and before it), they also resist the ways in which museums[16] and histories of LA perpetually fix Indigenous people in the past as vanished, disappearing, or extinct, instead of telling the vibrant stories of Indigenous Los Angeles that are part of the compelling and increasingly complex demographic diversity.[17] MILA portrays Indigenous communities in Los Angeles not as emerging but as the emerged groups who are often subsumed within other racial formations (Asian Pacific Islander [API] or Latino, for example). Even the term *American Indian* as an ethnocultural category covers the diverse ground of tribes that are federally recognized, those that are not, those that are native to California, and those that may have been displaced. Further, urbanization has encouraged new subcultures in which Indigenous people are often racialized within the Black, Chicano, Mexican, or Asian communities in which they reside. Other Indigenous communities are often erased, counted as Mexican or Guatemalan, or assumed to be Chicano or API, which does not account for mixed-race Indigenous peoples.

How do researchers, geographers, and community members work together to show the multiple stories of an untold Indigenous Los Angeles history? From the inception of the project, it was important to find a way to address all these issues and complexities of Indigenous Los Angeles outlined above in a way in which the narratives would not detract from one another but could still be in dialog. After exploring several possibilities, the ArcGIS StoryMaps platform was chosen by the MILA team as the primary visual storytelling forum for reasons that will be discussed in depth in this chapter. The platform provides a visual-spatial element, allowing for audiences engaging the material to contextualize the new information within the Los Angeles that they already know. Having a visual component is important in serving audiences that may be illiterate or able to read only English.[18] Using a website as the MILA project educational hub provides a place for all the maps to be featured together (figure 6.1). The curation of these digital maps allows viewers to engage the city in new ways. Participating communities or individuals represent themselves and their stories and communicate knowledge visually outside of the academic realm to a broader audience that navigates the world of web technology on a regular basis.

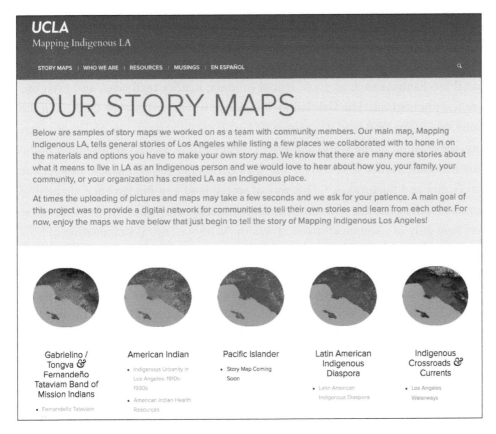

Figure 6.1. The MILA website's main page of stories (story maps).

Community-based mapping: The uses and drawbacks of GIS

The nature of this community-based project demands a critical examination of many of the basic assumptions of geographic histories and the implementation of colonial geographies. Otherwise, this project would be at risk of perpetuating many of the persistent and problematic narratives with origins in the period of colonization. In the case of the historic Los Angeles area, the colonial Spanish brought with them specific territorial-based economic and political organization models that were imposed on the landscape and its inhabitants in the late 18th century with the introduction of the missions and ranchos. The Spanish encomienda, a system of royal land grants and feudal labor, not only devastated the bodies of the original inhabitants of the Los Angeles area through violence, displacement, and slavery but also fundamentally changed the character of the landscape. This conflict of worldviews was maintained in the Spanish colonies and eventually became the dominant geographic narrative. That narrative is not isolated in history; it is maintained today through seemingly banal mechanisms such as grade school projects, history textbooks, and even theoretical foundations. The Gabrielino-Tongva, however, did not see the land in the same way.[19] The land (and its geographic locations) tells the stories of Tongva identity, the creation of the world and events through time, as well as providing all the raw materials needed for survival, but with these gifts was a responsibility for the land's care and nurturance.[20] This way of seeing land as a fundamental element of group identity contrasts with the Spanish colonial agenda which, in many ways, valued land only as a source of material wealth. Without critical examination of how geography has been used to dominate peoples, particularly as an extension of settler-colonialism, this project would inevitably reproduce many of those elements.

GIS is a catch-all term for many of the spatial and location-based technologies currently available. To say that GIS is just maps does not acknowledge the information element of these platforms. More to the point, an important function of GIS is to project data onto maps for the purpose of easy visual interpretation. This data tends to exist in spreadsheet format, and without the map, users could not easily make sense of what the data indicates spatially. Programs such as ArcGIS and QGIS are some of the premier tools used by researchers and professionals for such work. But in practice, these programs skew toward the physical: watershed analysis can draw and order complex stream networks, line-of-sight analysis determines what is visible from a discrete set of coordinates, and network analysis uses street networks for determining maximum service areas. These specific examples are emblematic of what GIS offers. It has revolutionized the spatial analysis processes for hydrology, telecommunications, and urban economics, respectively.

Indigenous place-based identity and perception

This project, however, raises many questions as to the usefulness of these platforms in telling stories of Indigenous place-based identity. The vast majority of professional GIS tools are almost exclusively designed for analysis, not storytelling or self-representation. Products produced in these platforms are static images more suitable for environmental impact reports or textbooks. This compatibility issue plagues GIS-based research in the social sciences, an issue that geographers continue to grapple with as the technology advances research in physical geography. ArcGIS StoryMaps, however, provides a platform with specific functionality that is considerably more useful for the purposes of this project. Dynamic, interactive maps engage the interests of the user while providing a creative outlet for representing place and identity. It is more user-friendly than the professional products discussed earlier, and the fact that this tool is free to the public more or less makes the project possible.

It must be acknowledged at the outset that the MILA storytelling platform is culturally biased in some significant ways. Particularly, the framework being used carries forward a colonial worldview that is not usually compatible, and often at odds, with worldviews of the Indigenous peoples. The top-down Web Mercator map projection on which most web mapping services rely is distinct from the ways other cultures have mapped their surroundings. It is a multiscalar digital evolution of the Mercator projection, which itself was developed in the 15th century for ease of transoceanic navigation from Europe.[21] It should not be forgotten that this technology played a crucial role in facilitating the colonization of the Americas. It is often taken for granted that these maps are not the only way to represent space. For example, sand paintings of the Gabrielino-Tongva are functional maps in a medium and spatial representation quite different from maps in the popular imagination (figure 6.2). This juxtaposition of spatial representations potentially puts the Gabrielino-Tongva community in the position of needing to remove culturally significant place-based stories from their original context and insert them into another. Furthermore, this sort of spatiocultural juxtaposition raises misleading questions about groups and their relationships to one another. Several maps currently on the

Figure 6.2. A Tongva sand painting.

Photo by Craig Torres.

internet draw boundary lines between some of the various Native groups of the Los Angeles area. However, in reality, territorial boundaries did not exist as such, and sharing of resources, languages, and family relations would have made such delineation unnecessary and not applicable. These basic assumptions built into modern-day mapping are pervasive and difficult to illuminate, as territorial delineation is essentially the sole purpose of today's political maps. In an effort to subvert some of these assumptions, the MILA team began exploring mapping styles that both fit onto the preset basemaps available from ArcGIS StoryMaps and break from these elements of territorial thinking.[22] One such mapping style involved depicting historic Tongva villages in Los Angeles by representing the relationships between historic Tongva villages: a network model superimposed on an Esri basemap.[23] While this approach is somewhat outside the box, it is data-friendly and relatively easy to generate.

TRANSLATING ETHNOGRAPHIC QUALITATIVE DATA

The digital basis of this platform necessitates that we translate stories and other memory-based history into data. Computer-aided mapping has caused a revolution in cartography and enabled widespread access to projects such as resilience mapping and community storytelling. Much of GIS is, however, reliant on data in a quantifiable format. In other words, the computer prefers numbers and logic. Several ideas discussed for the MILA project encountered the problem of data collection or translation. Scale quickly becomes a factor when dealing with data collected through interviews or survey. Historical records can be useful but always with the caveat that archival documentation is often biased and, in many ways, has been used to erase the relationships that have existed among communities and the landscape before colonization.[24] In cases in which recently aggregated data does exist, the cartographer must use creative methods for extracting the desired information. For example, there are shapefiles that map all the licensed restaurants in Los Angeles County. But if one is interested in Zapotec restaurants, how will they be identified from the total population? Furthermore, how will people account for the fact that only the formal economy is being acknowledged, thereby neglecting the informal economy? If a significant portion of that population exists in the informal economy, what use is the dataset in the first place? Data issues require a certain amount of translation, limiting which stories can be told and favoring those that can be quantified or represented as discrete phenomena in the landscape.

Ethical considerations in mapping

The professionalized nature of modern mapping and GIS can at times be at odds with the desires of the community, particularly when that professional relationship exists across differing cultural lines. As has become apparent over the course of the MILA project thus far, there can be differences of opinion between what should and should not be mapped, how specific location data should be, and the best way to represent that location data. Resilience mapping can be paradoxical; some communities that seek to assert their visibility in the landscape were targeted for removal or extermination in the past. How, then, can a community make its presence known in a spatial paradigm rooted in colonialism without providing the tools for those same institutions that have historically used knowledge as a method of cultural domination? It is then up to the mapmaker to map a location and people without necessarily making them fully visible. For example, abstracting the exact locations of certain phenomena is necessary to protect those who participate in the informal economy. This problem arose during the MILA project where street vendors in certain communities of Los Angeles would have been mapped. However, it was immediately clear that recording detailed location data could put those street vendors at risk of being targeted by law enforcement. Moreover, it is difficult to tell how consistent such locations are from one day to the next. Ultimately, the team agreed that heat map representations that fade out from a street or alley that is known to be a center for street vendors of a particular community strike an acceptable balance between exact and vague locations. In instances such as these, community sensitivities and professional limitations can be overcome through mutual understanding, collaboration, and innovation.

Working in Indigenous communities

An important aspect of working with Indigenous communities is ensuring that their voice is heard at each stage, from the choice of topics to the locations mapped. Too often, academic projects set early goals that may shape the direction of a project before the community joins the conversation. If the goal of a project at inception is to include community voices, the first step is to engage them to see if they want to participate or have suggestions such as alternate directions, goals, or participants. Equal footing and engagement build trust and create a successful project. We remain sensitive to this issue, specifically with respect to internal cultural protocols as well as to events or places being included that may be too traumatic or would compromise privileged information. By allowing community members to make decisions, the direction of the conversation may lead to unexpected topics or offer little-known histories that demonstrate the value of the project. This also allows for a first-person encounter for the map user and expands the value in a classroom or elsewhere outside of the community being represented.

There are several important issues around the responsibility of data gathering that continually arise in this project. The first concerns geotagging ceremonial locations or archaeological sites, which in some cases are legally protected information or may lead to harassment or intrusion during community activities. The California Public Records Act (Government Code Section 6254.10) exempts archaeological site locations from public disclosure to protect them from harm, such as looting or vandalism, and is part of the ethics of archaeology as a discipline. In this instance, the project shortened the geotag information to give a general location at a level that will provide enough specificity for the map for context but still protect the place from harm. Community members may have also chosen to do this for current ceremonial reasons or privileged locations following cultural protocols. This solution allows for a balance between sharing information publicly for education purposes and protecting privacy.

Another issue is recognizing the importance of temporal scales that situate people in the desired spatial atmosphere. It does no good to reference ranch land and waterways long ago demolished or cemented over with a dot on a modern map and no contextualizing information. Working with communities to consider time and space in reference to historical events is critical in helping users understand the place being discussed in a contemporary sense.

A final overarching issue to be raised is the recognition that these maps can provide information that may have political or even legal repercussions. For example, if current locations of places where undocumented laborers congregate are revealed, the information could be used against the very community being supported. However, thoughtful conversations with the community from a variety of vantages help to reveal potential areas of concern before the maps are created. This goes back to the importance of the community leading the conversation on what themes and places to include as it helps to avoid the potential pitfalls discussed.

Case study: Mapping Tongva LA

The Gabrielino-Tongva are Indigenous to and still inhabit the Los Angeles basin, though this often comes as a surprise to many who live in the city. As such, honoring Indigenous protocol and bringing visibility to the Gabrielino-Tongva as the original people of Los Angeles was the first task that the MILA project undertook (figure 6.3). Several important issues became apparent while doing this map and are important to discuss. First, since the Gabrielino-Tongva seldom have the opportunity to provide their own voices and experiences directly, it was imperative that it happen here. Second, there is a fine but crucial line to distinguish with the community to keep sacred or ceremonial sites private while inviting outsiders to share and learn. Third, communities never have one voice but reflect a diversity of lives, histories, and experiences of which expression is vital for a more accurate cultural representation. It was critical to the success of this project that we remember these issues for each of the communities and that we offer learning opportunities for this type of collaboration.

Figure 6.3. A Mapping Indigenous LA story showing a spring running through land maintained by the Gabrielino-Tongva Springs Foundation.

Photo by Mishuana Goeman.

We originally intended for this map to be largely self-directed and heavily based on community collaboration. The Tongva Los Angeles map would cover topics deemed by the community to be appropriate and important to share with the public. The MILA staff teamed up with two Gabrielino-Tongva educators, Craig Torres and Cindi Alvitre, to determine locations in the Gabrielino-Tongva landscape that should be included on the map. Torres and Alvitre provided great insight into each location chosen, as well as visual materials that had been collected over the years through Torres's own extensive research. What was not anticipated, however, was the time and effort necessary to transform this content into the correct format for input into the StoryMaps software. This task necessitated that someone on the MILA team essentially curate the information provided to fit the platform, as well as troubleshoot technology issues when they came up. Having an intermediary curator introduced a layer of editing and translation into the workflow that would ideally be avoided in endeavors of self-representation. Yet the reality of this project is that there are time-consuming back-end aspects that community members are not trained in nor generally have the time to do themselves. For example, community members have collected and contributed their own visual material to use in maps that we could not use until we obtained the proper permissions, even in cases in which their own ancestors were depicted. This task in archival research is extremely time-consuming. It remains an issue that needs to be resolved to remove the intermediary translation position and become more user-friendly and self-directed.

Many of the historical places important to the Tongva community were developed and are covered by urban infrastructure. Those that are still in existence often are vulnerable to vandalism or development. For example, educating the public about where Tongva villages were located throughout the city is important in seeing past the development and urbanization of the last century. However, exact locations of archaeological sites cannot be disclosed in order to keep them safe from looting. In addition, it is important to highlight that the Gabrielino-Tongva are still a vital and culturally vibrant community but are often spoken of in the past tense, as if extinct or culturally assimilated. While bringing visibility to contemporary Gabrielino-Tongva community activities and culture could negate this, ceremonies are generally solemn and private occasions, not for public spectators. Several solutions were identified during our meetings between MILA and community participants. To protect archaeological sites and ongoing ceremonial locations that were nonetheless important to discuss, we would not provide exact locations but keep the resolution general enough to serve the story. We also chose locations such as Kuruvungna (Serra Springs) where the public is invited to see cultural displays in the museum, visit the grounds to learn more, or attend a public event with the community.

The title, *Mapping Tongva Los Angeles*, is actually somewhat misleading. There is not one but many Tongva versions of Los Angeles, which is a concern that surfaced during work on this map. After working with Torres and Alvitre over several months to collect content for the map, including extensive video interviews, the team made a big push to get all the material into the application. At the first major feedback session with them, it became clear that this map was going in the wrong direction. The first draft of the map placed their voices in an uncomfortable position of authority over the landscape being depicted, when it was meant to be more about their individual perspectives. They also wanted more inclusion of the non-human, as is part of their cultural geographies. The use of video interviews woven throughout the length of the story could be divisive instead of informative depending on who was reading. Like any community, there is extensive diversity among the Gabrielino-Tongva. Views on the landscape, identity, and place differ on every scale and level, down to the family and individual. This point is important, as it is often assumed that Native communities are homogeneous in views, histories, and practice. It is necessary to ensure that multiple perspectives originate from within communities and not just between them. In response, we added more voices and faces to the video interview clips included to better represent the diversity and openness of the views about Tongva LA. The working title also changed to *Perspectives on Tongva LA* to be clearer. We also plan to keep an open format so that there is not one authoritative map. Working with communities cannot be rushed. This map took several years to finalize, but it achieves the goals set by community members in representing themselves and their stories.

Case study: Making a collaborative LAID map

Los Angeles is home to one of the largest Indigenous Latin American diasporic communities in the United States. Zapotecs, along with the growing diaspora of Mayans (largely Kanjobal and K'iche'), initially began migrating to the city in the 1980s as refugees of the widespread violence of the repression in Guatemalan Indigenous communities. According to community sources, there are over a million Guatemalans and Mexicans in the United States, many of whom are Indigenous.[25] However, their diversity is not distinguished, which this map is designed to correct. Floridalma Boj López, a contributor to the Latin American Indigenous Diaspora (LAID) site, said, "It is a way to make ourselves known, make ourselves present when demographic data doesn't do that for us. Some people give estimates, but I never trust that. I imagine it's much higher."[26]

The LAID map is designed to show how Indigenous migrants from Latin America continue to survive and make place through the maintenance and practice of spiritual and sacred beliefs as well as cultural, political, and social practices despite being thousands of miles away from their lands and across generations in Los Angeles.

As "people [are] displaced by militarism, neoliberal economic policies, and overlapping colonial histories," the LAID section of the project tells the stories of survival and being Indigenous in diaspora from the community themselves.[27] Sixteen places are represented in the LAID map (which can be viewed in full at www.arcgis.com/apps/MapTour/index.html?appid=31d1100e9a454f5c9b905f55b08c0d22), contributed by five researchers (figure 6.4).

Figure 6.4. The Latin American Indigenous Diasporas (LAID) page for the LA Comunidad Ixim youth group.

Being that the LAID map was a collaboration between several community contributors and representing several communities in one, the process raised some new issues that were not encountered in the first maps created in the project. For one, MILA had originally decided on using the Journal story template for all the maps because it provided the largest area for visual media and allowed us to maintain continuity across all four sections of the project. However, from the beginning of working on the LAID map, the contributors identified that this template was not going to work for their content. Because the Journal template works vertically, it inevitably creates a hierarchy in terms of information being presented, as something must be chosen to be featured first, next, and so on, until the last tab. This created an uncomfortable situation for the community contributors, who intended to represent their respective communities equally. The solution to this issue was to instead use the Map Tour template, which presents content horizontally and visualizes various places together on a shared map so the viewer sees them in relation to one another.

Zapotecs, Indigenous peoples from the Mexican state of Oaxaca, first began to arrive in the United States during the Bracero Program (1942–1964),[28] the recruitment of cheap Mexican male labor during World War II for agriculture and other labor-intensive production. However, it was not until the 1980s that Indigenous Oaxacans increasingly began to settle in the United States after the devaluation of the Mexican currency. There are two Zapotec groups mapped: Zapotecs from the Sierra Juárez (also known as Sierra Norte) and Zapotecs from the central valleys of Oaxaca. Due to their historical place of settlement and organization, Zapotec diasporas from the Central Valleys are mapped in West Los Angeles, while the Zapotecs from the Sierra Juárez are mapped east and south of Los Angeles proper.

Oftentimes, we read and hear more about Zapotec peoples living in South and East Los Angeles, which not only contributes to the invisibility internally among Zapotec groups but also homogenizes their experience and ways of knowing their worlds. Currently, there are three maps (St. Anne's Church in Santa Monica, California; Guish Bac Folklorico; and COTLA: Comunidad Tlacolulense en Los Angeles) that tell the stories of valley Zapotec as they relate to religious and sacred beliefs, and cultural and political practices in West LA. The map shows the ways in which these ceremonies and events take place as they bring the diasporic communities together. For valley Zapotecs in West Los Angeles, faith-based processions on the city streets with the statue of their village's saints, incense, flowers, an Oaxacan brass band, and the singing of traditional and colonial prayers are yearly occurrences that become visible in the Latin American Indigenous Diaspora part of the MILA project.

For representation of the Zapotecs from the Sierra Juárez, The Centro Educativo Benito Juárez (CEBJ), Zoochina: Second-Generation, and DJ Survive from Zoogocho were included. Each of these stories is accompanied with a video. What it shows is how the United States-raised Zapotec generation continues to maintain and reclaim their traditions and sense of belonging to their community in diaspora, whether it be the town of Yatzachi el Bajo, Zoochina, or Zoogocho, which respectively correspond to the mapping sites at the beginning of this paragraph. Each individual shares their unique way of belonging to their community and the reasons they have maintained connections even across generations in diaspora. At the same time, the stories also attempt to highlight their professional careers to diverge from historically racist stereotypes that Indigenous diasporas are lazy or can only work in blue-collar jobs.

Indigenous Guatemalans and Oaxacans increasingly migrated in the 1980s and established rich transnational communities. As observed from the Maya (LA Comunidad Ixim, Pastoral Maya of Los Angeles, Mayavision, Manuel Jaminez Xum) and Zapotec portions of the map, there are second- and third-generation youth who refuse to lose their roots and have become active in their community's cultural revitalization. Many service providers have failed to account for the population of recent immigrants who are often non-Spanish monolingual speakers of their own native languages. These communities are different from their Mexican or Guatemalan counterparts. The LAID map captures both the tragedy and longevity necessary to survive across diasporic generations.

Being that the LAID map was a collaboration between several community contributors and represents several communities in one format, the process raised new issues that were not encountered in the first map created for the project. In response to this process, MILA has since encouraged the use of any of the StoryMap templates available in recognition that for different community content, different spatial configurations of the content itself may be appropriate. For instance, our pow-wow map is organized horizontally by seasons. Creative conversation with community contributors and researchers has given the project insight into differences in priorities and preferences when using StoryMaps, such as gaining more even representation in exchange for larger visuals and media.

Case study: Urban Native LA

From the beginning of this project, one request that the MILA team received on several occasions was to create a map of resources that serve Native communities in this large and diverse region. This task is important for several reasons. Practically speaking, health, education, and community support services that cater specifically to American Indian or Indigenous constituencies are often difficult to locate. Similarly, outreach on the part of the organizations that offer these services is challenging. In a city with the largest population of urban Indigenous peoples in the United States, including the original inhabitants of the Los Angeles area as well as those who have relocated here, these locations often serve as important sites for community gathering and place making—a theme central to the MILA project. Bringing visibility to valuable resources for necessities such as health care, childcare, family support, and educational support simultaneously shows Native Americans where they can find community across a vast geography.

To compile the data to include in the Urban Native Los Angeles resource maps, MILA worked with several undergraduate and graduate students at UCLA who come from and often work in these communities. Undergraduate students took on the task of researching services offered to Native peoples in Los Angeles, noting locations, contact information, and the types of services offered. Graduate students did more in-depth research about the history of services in the city, and at times included their relatives' personal journeys to ground the maps (figure 6.5). Faculty worked weekly with undergraduates and biweekly to develop material. The students then worked with MILA project staff to complete the resource maps. We held large community workshops every three months to help upload the material.

Figure 6.5. American Indian Health Resources web mapping application.

Photo by Mishuana Goeman.

Because of the changing nature of the data, the Urban Native LA resource maps will likely always be a work in progress. Contact information and services offered need to be kept up to date for the maps to remain useful over time. To solve this issue and broaden the information provided, MILA staff will work with these organizations to create their own maps, as opposed to being part of a static list with a minimal spatial element. As an example, United American Indian Involvement (UAII) would like to create a map in which the organization can chronicle its history and important places and use our project for outreach to larger communities. Direct representation from these organizations will ensure that their voices and goals are self-represented and current. This collaboration will provide much more useful information to the public and community.

FINAL DISCUSSION

Using ArcGIS StoryMaps as the main environment for MILA community content has had both upsides and downsides. The nature of the stand-alone environment affects how useful the products can be for geospatial representation. For example, we wanted users, particularly those with limited computer experience, to interface with the program directly by adding map notes. Instead, we decided to create layers in ArcGIS and import them into ArcGIS Online. Later, data was compiled into primary spreadsheets that were uploaded to StoryMaps directly. While this process in StoryMaps is simplified from the process necessary to convert spreadsheet data into a shapefile using ArcMap™ (for example, it is not necessary to geocode addresses to coordinates in StoryMaps), it still presents a technological challenge for many users. As the project continued to move forward, accessibility and simplicity had to be tempered by the way the program accepts data. The limited ability to make technical maps for geospatial analysis is not necessarily an issue for everyone. Given that map layers can be uploaded into ArcGIS Online, those who seek this level of detail in their map may still do so. For many other users who create maps, having basic mapping capabilities is sufficient.

One of the main reasons MILA was drawn to this program was its user-friendly nature. The project needed software that allows users to make their own maps with little assistance. The idea was to create short written tutorials detailing how to get started, as well as to provide additional support via email. For many students making simple stories and resource maps, this program has been successful. However, we have found that there are situations in which the user-friendly factor is not enough, such as working collaboratively with several people on one map or working with community members with limited ArcGIS skills. In some of these cases in which more support is necessary, a staff member or researcher has taken over the map building as a point person, editor, and curator of the content, as described in some of the case studies.

An important feature of StoryMaps is that any number of community members and researchers can generate products at the same time on their own or with assistance. As communities change and expand, the platform remains accessible. As maps are completed and ready to go live, a link to a map is added to our website, and the maps become part of an assortment of stories and perspectives about Indigenous Los Angeles. The ability for community members or researchers to edit their stories as necessary without MILA needing to update the website also goes a long way in keeping information current. On a similar note, StoryMaps is conducive to the goals of self-representation and first-person storytelling within the project. In addition, the ability to combine text and visual components such as images and videos with a spatial component is important in bringing cultural and historical depth to the familiar sights of contemporary urban Los Angeles. Hidden, invisible, or erased stories that often remain only in memory among small groups of people are visualized when they are embedded or reasserted back into the landscapes presented to the public on our website.

One unresolved concern that seems to be a problem for many public cultural projects is the issue of data hosting. Using a third party to host content can revoke sole ownership of culturally sensitive material from the community. Through easily dismissible terms of service agreements, communities may find their cultural property appropriated for the benefit of some commercial interest. Potential for appropriations notwithstanding, inheritors of cultural property are wary of their content being used irresponsibly or out of context in ways that could be detrimental to the community. This predicament extends to any faculty, principal investigators, or researchers and their discretion on how to treat data used by the project. Therefore, being transparent about this to people sharing their content is important. Ideally, content shared is ultimately screened by the community members as being appropriate for public consumption from the start. Further, MILA is intended to be a resource for K–12 education as well, and maps are created and included in the project with this in mind.

Going forward, MILA has identified several new directions in which to take the project. One feature that is not currently available that would make historical cultural storytelling more relevant to communities would be the ability to replace the preloaded basemap with a referenced or unreferenced historical map. By enabling the user to upload a picture of, say, a sand painting and build a guided tour through that medium, exciting opportunities for cross-cultural exchange are possible that do not require translating one geographic worldview into another. Were this feature made available, storytelling capabilities in differing conceptions of landscapes would be much broader.

Another idea that MILA intends to pursue is to create a mobile app in addition to the content on the website. Having an app that users could interact with on location at various sites in Los Angeles would increase the impact of the immersive experience. An app would also add a new layer of crowdsourced data collection, which could go a long way in battling the issue of often having low amounts of data.

CONCLUSION

Given the many challenges presented in making spatial representation available to Indigenous communities of Los Angeles, the participants of this project are hopeful that this will become a widely used resource available to a public audience that may not be aware of the nuances of the city in which they live. Countering problematic narratives that valorize the mission system, bringing forward ancestral stories, providing a list of resources available for community members, and bringing visibility to that which was erased may seem ambitious for a single project, but it is the collaborative nature of this project that makes such a wide range of goals possible. ArcGIS StoryMaps has provided a communal working environment in which to facilitate such collaboration by presenting a standardized toolset and shared database. Despite the challenges encountered over the course of the project, the fact that it can be undertaken at all is in many ways made possible by StoryMaps and its widespread availability.

Studying Indigenous Los Angeles is a way to study the multiple regions that meet in Los Angeles—the Pacific, Latin America, Native America, and Native California—from an unexpected point of view, thereby allowing us to rethink the settled and congealing assumptions about the city and the diversity of its inhabitants. Using indigeneity as a window and point of departure for the study of LA, we can understand racialization in a multiethnic/multicultural environment in a new, innovative way that is grounded in the original peoples of now-California and the Indigenous diasporas that have come to populate this land. While most of these communities have been displaced by war, militarism, governmental policy, and global economic schemes, we rarely consider them in relation to one another. No other project exists that focuses on comparative indigeneity and comparative racialization in an urban center. This is in part because no other place exists like Los Angeles. The MILA project provides an important platform to document the interconnected histories of Indigenous LA as well as raise the question of what it means to reaffirm identity and make place in another Indigenous group's land.

REFERENCES

Avila, Eric. 2006. *Popular Culture in the Age of White Flight: Fear and Fantasy in Suburban Los Angeles*. Berkeley, CA: University of California Press.

Cavanagh, Edward, and Lorenzo Veracini. Settler Colonial Studies Blog. 2010. https://settlercolonialstudies.org/about-this-blog.

Dillow, Gordon. 1993. "No Common Ground: Indians Say 22-Acre Site Is Sacred but Cal State Long Beach Disagrees and Plans to Lease It to Developer." *Los Angeles Times*, June 19, 1993.

Ellinghaus, Katherine. 2014. "Mixed Descent Indian Identity and Assimilation Policy." In *Native Diasporas, Indigenous Identities, and Settler Colonialism in the Americas*, edited by Gregory D. Smithers and Brooke N. Newman, 297–316. Lincoln, NE: University of Nebraska Press.

Empowering Pacific Islander Communities. 2014. *A Community of Contrasts in California*, 2014, 64 pp., Los Angeles, CA.

Fixico, Donald. 1990. *Termination and Relocation: Federal Indian Policy, 1945-1960*. Albuquerque, NM: University of New Mexico Press.

Goeman, Mishuana. 2015. "Land as Life: Unsettling the Logics of Containment." In *Native Studies Keywords*, edited by Stephanie Nohelani Teves, Michelle Raheja, Andrea Smith, 71–89, Updated Reprint. Tucson, AZ: University of Arizona Press.

Jurmain, Claudia, and William McCawley. 2009. *O, My Ancestor: Recognition and Renewal for the Gabrielino-Tongva People of the Los Angeles Area*. Berkeley, CA: Heyday Books.

King, Chester. 2011. *Overview of the History of American Indians in the Santa Monica Mountains*. Unpublished Manuscript, October 2011. Microsoft Word file posted on Academia.edu.

Monmonier, Mark. 2004. *Rhumb Lines and Map Wars: A Social History of the Mercator Projection*. Chicago, IL: University of Chicago Press.

Norris, Tina, Paula L. Vines, and Elizabeth M. Hoeffel. 2012. *The American Indian and Alaska Native Population: 2010*, US Census Bureau, 21 pp., Washington DC.

Ramirez, Renya. 2007. *Native Hubs: Culture, Community, and Belonging in Silicon Valley and Beyond*. Durham, NC: Duke University Press.

Rosenthal, Nicolas G. 2012. *Reimagining Indian Country: Native American Migration and Identity in Twentieth-Century Los Angeles.* Chapel Hill, NC: University of North Carolina Press.

Saldaña-Portillo, María Josefina. 2016. *Indian Given: Racial Geographies across Mexico and the United States.* Durham, NC: Duke University Press.

Stephen, Lynn. 2007. *Transborder Lives: Indigenous Oaxacans in Mexico, California, and Oregon.* Durham, NC: Duke University Press.

Stoler, Ann Laura. 2002. "Colonial Archives and the Arts of Governance." *Archival Science* 2 (2002): 87–109.

Wolfe, Patrick. 2006. "Settler Colonialism and the Elimination of the Native." *Journal of Genocide Research* 8 (2006): 387–409.

NOTES

1   Norris, Vines, and Hoeffel, *The American Indian*, 11. According to the report cited here, US census data still counts New York, New York, as the largest American Indian population. However, if one includes the diasporic Indigenous peoples from Latin America and Oceania to the cited 54,236 American Indians in the city, Los Angeles far surpasses the 111,749 American Indians of New York. Numbers for diasporic Indigenous peoples from Latin America and Oceania are discussed further in this chapter.

2   Settler colonialism is defined by Edward Cavanagh and Lorenzo Veracini in their blog, Settler Colonial Studies Blog (2010), as "... a global and transnational phenomenon, and as much of a thing of the past as a thing of the present. There is no such thing as neo-settler colonialism or post-settler colonialism because settler colonialism is a resilient formation that rarely ends. Not all migrants are settlers; as Patrick Wolfe has noted, settlers come to stay. They are founders of political orders who carry with them a distinct sovereign capacity. And settler colonialism is not colonialism: settlers want Indigenous people to vanish (but can make use of their labour before they are made to disappear). Sometimes settler colonial forms operate within colonial ones, sometimes they subvert them, sometimes they replace them. But even if colonialism and settler colonialism interpenetrate and overlap, they remain separate as they co-define each other." See also Patrick Wolfe's explanation of the term understood through the logic of elimination in "Settler Colonialism and the Elimination of the Native."

3   Ellinghaus, "Mixed Descent Indian Identity." The article defines *diaspora* as "... the connections a community of people maintain over distance."

4   We use the word *Native* only to refer to American Indians, relocated or not. We use the word *Indigenous* to refer to a collective group, including the diasporic and Native communities found in Los Angeles.

5   United States Census Bureau 2016 Population Estimates by Metro Area.

6   Avila, *Popular Culture in the Age of White Flight*, 34–41.

7   The MILA project was first funded with $20,000 in 2013 by the UCLA Institute of American Cultures Dream Fund Research Grant.

8   Dillow, "No Common Ground." See also Loewe, Ronald. *Of Sacred Lands and Strip Malls: The Battle for Puvungna* (Lanham, MD: Rowman and Littlefield 2016).

9   Indian Relocation Act of 1956, Pub. L. 959, 70 Stat. 986 (1956). The name for this era, "relocation days," as it is called by many who remember it, is derived from the name of the policy itself.

10 Nicolas Rosenthal looks extensively at the networks built between cities, towns, the rural, and reservations during the relocation era, taking Los Angeles as a focal point. See Rosenthal, Reimagining Indian Country. See also Fixico, Termination and Relocation, 134-157.

11 In Native Hubs, Renya Ramirez uses the term *transnational hub* in the context of American Indian and Indigenous diasporic communities in the United States.

12 *Latin American Indigenous Diasporas.* Mapping Indigenous Los Angeles, accessed July 29, 2016 at www.arcgis.com/apps/MapJournal/index. html?appid=a9e370db955a45ba99c52fb31f31f1fc. Indigenous Oaxacan organizations in Los Angeles estimate that there are 120,000 Indigenous Oaxacan immigrants living in the city.

13 LA Comunidad Ixim; Pastoral Maya of Los Angeles; Mayavision; Frente Indígena de Organizaciones Binacionales (Indigenous Front of Binational Organizations); Guish-Bac Folklorico; Comunidad Tlacolulense en Los Angeles (COTLA); and Centro Educativo Benito Juarez (CEBJ).

14 "Empowering Pacific Islander Communities and Asian Americans Advancing Justice," A Community of Contrasts in California, 2014, 7.

15 "Empowering Pacific Islander Communities," 38.

16 See MILA's blog about the Los Angeles Natural History Museum.

17 Jurmain and McCawley, *O, My Ancestor*, 7.

18 Respectively to each site, the project includes video clips in English or Spanish. Currently, the team is working on providing English subtitles to its non-English videos.

19   For an example of contemporary perspectives, see MILA's map *Perspectives on a Selection of Gabrielino/Tongva Places.*

20   Jurmain and McCawley, *O, My Ancestor,* 103.

21   Monmonier, *Rhumb Lines and Map Wars,* 6.

22   Goeman, "Land as Life," 71–89.

23   King, "Overview of the History of American Indians in the Santa Monica Mountains." This concept was inspired by a similar network model of Native Southern California by Chester King.

24   Stoler, "Colonial Archives," 87–109.

25   There is no official way to track who is Maya or Zapotec, which is why this mapping project is important. The census does not account for Mayas or any other Indigenous group.

26   Boj López, personal communication, December 4, 2015. At the time, Boj López was a doctoral candidate in USC's Department of American Studies and Ethnicity.

27   Research Scope, Mapping Indigenous Los Angeles, accessed July 29, 2016. https://mila.ss.ucla.edu/636-2/research-scope.

28   Stephen, Transborder Lives. See also Saldaña-Portillo, Indian Given.

→ CHAPTER 7

Indigenous Martu knowledge: Mapping place through song and story

SUE DAVENPORT AND PETER JOHNSON

INTRODUCTION

This story is about how one community has worked to rebuild social stability and resilience[1] through a combination of ancient and modern technologies and knowledge. It highlights ways in which two cultures understand country and how melding those understandings can create strength and opportunity.

Aboriginal people have lived in the Western Desert of Australia for millennia (Veth 1989, 116). Martu are among the most traditional of Aboriginal groups in Australia, with many of the older people leading a completely traditional desert life, devoid of knowledge of the Western world. In this chapter, we look at ways in which geography and Martu knowledge systems built around geography were integral to their traditional life and are now helping them to deal with their modern context.

We begin with the ecological, cultural, and social background of the Martu. We then explore the fundamental nature of geography to the organization of knowledge in this traditional society. We argue for an expansive interpretation of the term *geographic information systems*, to recognize the richness and potential of geographically based organization of information. Finally, we examine the practical ways in which Martu people are using geography and geographic information systems in reshaping their traditional society to meet the challenges of their postcolonial situation.

MARTU

Since their native title determination (*James v. Western Australia,* 2002), Martu people hold title over 50,000 square miles of desert country in the Great Sandy, Little Sandy, and Gibson deserts of Western Australia (figure 7.1). Their country has been described by American anthropologist Richard A. Gould (1969, 273) as "the harshest physical environment on earth ever inhabited by man before the Industrial Revolution."

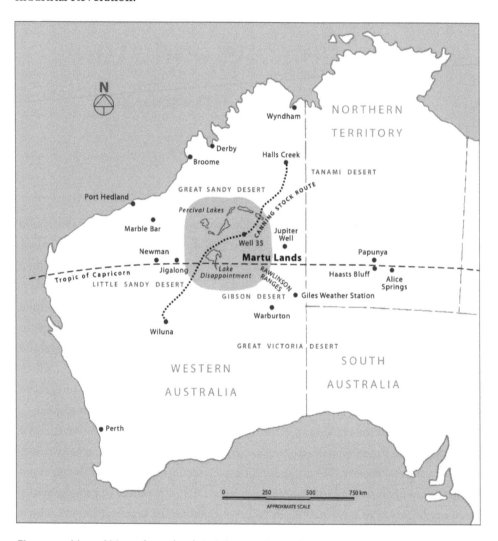

Figure 7.1. Map of Martu homelands in Western Australia.

Map courtesy of Kanyirninpa Jukurrpa.

Martu are an amalgam of traditional groups. They speak a variety of dialects and have belonged to a variety of landscapes within the vast area over which they now hold title. While they do not have a uniform history nor completely uniform culture, they identify as an integrated group in their modern situation. Their dialects belong to a common family, and many of the foundations of their societies were consistent, if not identical. For our purposes in this chapter, we refer to them at a high level as a single cultural bloc.[2]

Martu have continuously occupied their desert country for thousands of years. Radiocarbon dates from five stratified caves and rock shelters in the Karlamilyi (Rudall) region illustrate continuous human occupation from at least 5,000 years ago. (Veth 1989, 116)

Many Martu (figure 7.2) remained out of contact with the Western world until the second half of the 20th century (Davenport, Johnson, and Yuwali 2005, 154). Families came in from the desert to facilities run by European Australians ("whitefellas" to Martu) in two waves, initially through the 1930s and 1940s and then accelerating through the 1950s and 1960s. Their country was seen as essentially worthless by the Western world until the 1980s, when a small number of mineral deposits were discovered and began to be developed. Throughout most of Australia's colonial history, Martu were thus substantially, if not wholly, left alone.

Figure 7.2. Martu leader Nyari Morgan, 1963.

Photo courtesy of T. Long.

Martu engagement with the Western world has been limited and distinctive. From the 1940s until the 1970s, it was chiefly engagement with small, remote, Christian missions and vast, isolated pastoral stations (Tonkinson 1991). The contact was therefore with limited and even idiosyncratic subsets of Western society—zealous missionaries or a scattering of single men who managed very remote stations. From the 1970s until now, most Martu continue to have limited engagement with the West, having returned to their desert homeland and set up several remote communities in which their children grow up.

What has changed and what has remained?

Modern life in these communities is manifestly different from traditional life. Martu lead a more sedentary lifestyle, no longer walking their country as hunter-gatherers as they did traditionally. Many have returned to live in the desert homelands over which they hold native title, in communities located in the country that they walked traditionally. They have access to Western technologies: housing, electricity, vehicles, guns, fast food, alcohol, mass media, and medicine. On the whole, they have adapted these to their social modes and preferences, but their assimilation of Western resources into their society has had its costs: new phenomena of substance abuse, chronic lifestyle diseases, early adult mortality, crime, and incarceration.

On the other hand, they manifestly retain many of the deeper systems and structures from traditional life. Their cosmology, language, religion, ceremony, kinship-based social structures, sources of social status, motivations and aspirations, foci of respect and reverence, social priorities, and lore all derive from their traditional culture. Their identity is trenchantly Martu, rooted in their traditional society.

The foundation of Martu identity, culture, and society is country. At a profound level, individuals, families, and groups belong to specific sites and tracts of country. Geography provides the template for organizing much of their cultural knowledge.

MARTU KNOWLEDGE SYSTEMS

The Martu world is ordered by *jukurrpa*.[3] This word, which has no English equivalent, covers several meanings: the creative epoch, during which the land, the law, and the people all were given shape; the stories and songs that recount that time—the travels and adventures of the creative beings; the eternal, immanent power that derives from the creative beings; the continuing, vital action of that immanent power; and the ceremonies and sacred objects that represent, hold, and engage with that eternal power (Tonkinson 1991). Where an English equivalent is sought, *jukurrpa* is usually translated as *dreaming*.

> My dreaming is there to the west. It was that dreaming which left me there,
> left my spirit in the ground there to enter my mother and be born. It was
> the Warinykuriny dreaming. It is the dreaming from over there in the West
> at Katakurtuji, and Kurunyuriny. (Taylor and Muuki 2009)

Martu *jukurrpa* is largely held and communicated in songs and stories (figures 7.3 and 7.4). These provide mnemonics for the storage and retrieval of vast storehouses of information. Some stories and songs are open—able to be learned by children. Many are restricted, holding knowledge accessible only to appropriately inducted men or women. Over a lifetime, a Martu elder may have learned vast song cycles, many thousands of lines containing a layered revelation of knowledge, as well as a storehouse of more detailed stories. This accumulated knowledge is termed *ninti* and regarded with great respect. Through knowledge, an elder gains both social status and the power to engage with or hold off supernatural forces.

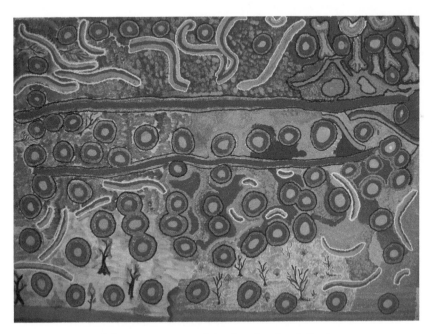

Figure 7.3. Map of the country east of the Canning Stock Route.

Image courtesy of Kumpaya Girgiba, artist.

Singing the country

The songs and stories speak of country: of the identification, source, and character of specific geographic features. They chiefly concern the activities of myriad characters from the creative epoch who traveled the country, meeting each other, having adventures, and forming geographic features as they went. Lakes, rivers, creeks, hills, soaks, and other features are seen as manifestations of the activity of these creative beings.

The *songline* traces the route taken by dreaming travelers and dramatizes, albeit cryptically, their activities, both notable and mundane. Songlines are a network of routes, sites, and landscapes that traverse the thousands of square kilometers of Martu country. They are also a celebration, in lyrical and dramatic form, of the interplay of land and the forces of nature with dreaming powers (Tonkinson 1991, 124).

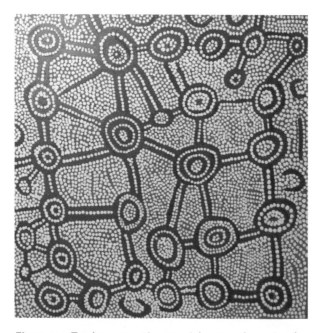

Figure 7.4. Tracks connecting special water places for the *Jakulyukulyu* Seven Sisters. Image courtesy of Nyari Morgan, artist.

Precise information can be hidden from Western eyes and ears but is immediately apparent to Martu:

Then he (ancestral being) went back to his camp and got those two women (his wives) and took them. He kept on going with them right up to

Nyukuwarta. They stayed there. They are there today (three rocks standing together). Those women are at Nyukuwarta. Those are the Nyiminjarra dreaming men. (Williams 2009)

The lines of the dreaming travelers form songlines, which continue over great distances into neighboring country. When a song moves from one country into another, the song changes to the appropriate language.

One women's songline, which Martu call *Jakulyukulyu*, tells the story of the Seven Sisters (figure 7.5) and runs from the east coast to the Western Australian coast near Roebourne and back into Central Australia (James 2005, 33).

Figure 7.5. *Jakulyukulyu* Seven Sisters.

Image courtesy of Muuki Taylor, artist.

Songlines form a tapestry covering the continent, made up of interconnected poetry in different dialects and languages, entirely held in human memory, distributed across the land. They represent a distinctive canon of world literature, solely orally transmitted, largely unknown to and unrecognized by the Western world.

The creative beings were supernatural, eternal. They thus still exist: having finished their creative endeavors, they entered the ground or flew into the night sky. A waterhole may today be the home of a rain-making snake whose travels are complete but who is still present and powerful. The country is literally alive—not simply with flora and fauna, but with eternal, supernatural life. Sites are identified not merely as representative of *jukurrpa* characters, but as manifestations:

> This black water snake belongs to us, it is ours. From a long time back the ancestors looked after this *Jukurrpa* dreaming for a long time and then they died and they left it to us. We are looking after it. That is all. This is our black snake permanent water. (Taylor and Waka 2009)

Mapping country

The songs and stories that speak of *jukurrpa* provide rich intellectual maps of country.

> Through the learning of songlines, which contain both geographical and mythical referents, Martu are familiarized with many hundreds of sites they may never have visited, yet all become part of their mental map of the desert world. Because ancestral beings traveled from waterhole to waterhole much of the time, song sequences imprint knowledge that may have practical value should people be traveling in the area depicted (Tonkinson 1991, 124).

In describing the activities of the dreaming beings, the stories and songs include practical information critical to survival in this harsh environment: the location of water sources; locations and descriptions of significant domestic and ceremonial sites; the key topographic markers of the routes between water sources; knowledge of species, their locations, and habitats; astronomy; seasons and their relationship to species propagation; and the spiritual nature and significance of a place, including the dangers or powers associated with a site.

> The Tjukurrpa, as geographic narrative and atlas, penetrates the mysteries of ancient desert survival in arid glacial times. It alone counts the names, codes, and descriptions of all the waterholes—all the places where survival was possible—across the entire arid zones (and the greater Australian continent). The Tjukurrpa thus provides the ultimate survival guide in the context of environmental extremes and extreme glacial aridity. It is, in this sense, the last word on the matter (Cane 2013, 167–168).

Mapping society

Songs also include information of social significance: the foundations of the kinship system; the foundations of language and its relationship to location; the rights to use country or waterholes; the allocation of different country to different languages; the "mix-up" places where groups come together; the myriad rules governing social interaction and behavior; the forms of social etiquette; the laws that govern social behavior and the punishment of transgressors; and the authority to punish.

The songs hold the foundations of Martu spirituality: the origins of the universe; the origins of groups of people; the identity and characteristics of different groups of people; the location, threats, and dangers of different supernatural beings; the form and content of ceremony; and the passage of men and women through stages of revelation and lore.

> The imperatives embodied in the Dreaming are the Law, which situates the origin and the ultimate control of power outside human society; that is as emanating from the withdrawn but still watchful creative beings of the spiritual realm. From their human "descendants," these beings demand conformity to the Law and the proper performance of ritual. In return they will ensure the reproduction of earthly society through a continuing release of life-giving power into the physical and human world. (Tonkinson 1991, 106)

Accessing knowledge of country

We once accompanied an old man as he traveled through the desert. He was looking at a map of the desert and pointed out the single desert track marked on the map. "Whitefellas only have one line here," he said. He then drew a complex grid of lines in the sand. "This is a Martu map—lines everywhere." The map is not usually drawn; it is held in memory, in song (figure 7.6).

These songs are rooted in landscape in a way that allows people who have never been to a location to navigate to it, to recognize it immediately, and to identify its significance and restrictions. We once traveled with an old man to a waterhole far out of his country. When asked whether the site was a women's place, he was unable to answer immediately. Overnight, he sang the site and in the morning identified it—its features, its religious significance, and its place in a travel route.

Figure 7.6. A map of grid lines in the sand, drawn by a Martu old man.

Photo courtesy of Sue Davenport.

Another time, we traveled with a group of old men, who came for the first time to a site with a vast gallery of rock art. On seeing the petroglyphs, they spontaneously broke into song, telling the story of the site. They knew the place intimately, despite having never been there.

This spontaneous breaking into song is commonplace as people travel through the country. The country is a narrative, which they read as they travel. Or, rather, it is an interwoven tapestry of stories, linking at key sites, each story following its own distinctive path through different parts of the country. One song will run north, another west. One will follow a river system east, another a track south through the desert. At any site of significance, the song can be picked up, but it is not an isolated fragment. It is a segment from a continuous canon that covers the landscape. So, as people travel through the country, they are inside the songs, surrounded by them.

Country and identity

Martu believe that they are manifestations of *jukurrpa*, their spirit having resided in the country since the time of the creative beings. At the time of their conception, they acquire a conception totem (*jarriny*) from the country, flora or fauna.

The place from which a person's spirit comes is his or her Dreaming place, and the person is an incarnation of the ancestor who made the place. A person's Dreaming provides the basic source of his or her identity, an identity that pre-exists. (Myers 1986, 50)[4]

My two dreamings took me and made me strong in the dreaming and put my spirit to be at Warnpurrtinya, at the place Warnpurrtinya. My spirit stayed there and then I was born. After being born I went—my parents carried me around as a small child. They carried me round as a baby around my country Warnpurrtinya and around Puntujarrpa; they took me around there. (Bennett 1987)

Martu, individually and collectively, source their identity in country. Individuals belong to specific sites and tracts of country through a variety of connectors: conception totems, birthplace, parents' or grandparents' identity with country, language, marriage, social induction as a countryman, site of a parent's or grandparent's death, or custodianship of a particular songline segment. The country does not belong to them—at a social and a metaphysical level, they form part of and belong to their country.

Country and belonging

Imagine yourself naked, alone in a vast, harsh landscape. The immediate Western response is a sense of threat—something we see regularly when Western people enter this landscape, and something we would feel if we were isolated, but which is allayed when in the company of Martu. Our sense is that, for Martu, this existential sense of threat is eased by a combination of three factors. Martu do not stand outside as an interloper in the country, as Western visitors do. They are, in a profound sense, part of the country—they belong, indivisible from country. Then, when traveling in their own country, they have intensive, intimate knowledge—of sites and travel routes, of water sources, of weather, of flora and fauna. This practical knowledge neutralizes most natural, practical concerns, in all but the most inhospitable times. Finally, elders, through their esoteric knowledge of country and the supernatural threats it contained, have the power to ward off those threats. Through that knowledge, they participate directly in the forms of power that course through the living body of the country.

A camp can be made almost anywhere within a few minutes—a wind-break set up, fires built, and perhaps a billycan of tea prepared [figure 7.7]. Unmarked and wild country becomes a "camp" *ngurra* with the comfort of home. The way of thinking that enables a people to make a camp almost anywhere they happen to be, with little sense of dislocation, is a way of thinking that creates a universe of meanings around the mythologized country. (Myers 1986, 54)

Figure 7.7. Martu asleep in their camp at Well 29, Canning Stock Route, 1964.

Photo courtesy of T. Long.

When in their country, Martu had a geographically ordered and encompassing system of knowledge. This allowed them to belong within, make use of, and cooperate with their habitat. They had intensive knowledge of a specific landscape, its features, resources, dangers, and behaviors, rather than the modern, general Western knowledge of a planet. Their intensive knowledge was their technology: for navigation, for survival, for social organization, for spirituality. That knowledge derived directly from the forces that shaped both the country and them: *jukurrpa*.

In synthesis, amongst Martu, complex philosophical concepts associated with life essence or spirit (*kurrurnpa*), the perpetuation of that spirit through species in the landscape, the emergence of spirit-children (*jarrinypa*), "owning" [or] holding (*kanyirninpa*), knowing and teaching (*ninti-*)

about country, sharing country with kin (*waljamarra*), and more were tightly integrated concepts, at least in traditionally orientated contexts or idealized scenarios. (Walsh 2008, 307)

GEOGRAPHIC INFORMATION SYSTEMS

Is such an encompassing, geographically based knowledge system, held in memory and reliably passed down from one generation to the next, a geographic information system (GIS)? In this section, we argue that such an expansive conception of GIS, such an extension of discourse about GIS, leads to a better understanding of the potential of GIS technologies to contribute to the richness, resilience, and sustainability of societies.

An understanding of the depth at which geography is imprinted on Indigenous knowledge, providing the organizing principle for traditional knowledge, provides an insight into the proper foundations for all manner of modern, social initiatives for Indigenous communities. This is demonstrated in later sections of this chapter, which discuss the extent to which modern GIS software and GIS-assisted technologies are melding with traditional knowledge and techniques in an array of social, environmental, and cultural programs in Martu communities.

The mainstream world uses technologies to capture, preserve, and use knowledge of landscapes. GIS software provides a template for geographically based organization, analysis, and retrieval of all sorts of information. Precise and powerful systems can be harnessed in imaginative ways to gain knowledge and insights that previously may otherwise have been hidden. The range of existing and potential applications in scientific, industrial, and social spheres is vast. Geography provides a credible and useful organizing principle for all manner of knowledge.

GIS usually refers to Western conceptions of data (measurable, quantifiable) and Western forms of technology. Within conventional contexts, this limited view of GIS is legitimate.

Within our domain, this limited view of GIS has three drawbacks. First, it limits our understanding and recognition of human genius. An expansive view would engage with the richness of a culture that organizes almost all its knowledge, its literature, and its social and spiritual foundations on the basis of geographic keys—a culture in which comprehensive knowledge of a landscape grounds identity and existence. This ought to be of more than passing interest. An appreciation of the complexity and beauty of such a knowledge system provides insights into the breadth of humanity, human cognition, and ingenuity.

Secondly, it leads to lost opportunities to engage effectively with Indigenous societies in transition and to bolster and harness their social resilience. In Australia, the nature and direction of that transition and the failure of Western agencies to provide effective support to that transition are widely recognized as sources of national shame.[5] When an Indigenous society is so fundamentally grounded in geography, one would have thought that policy responses to its aspirations, adaptations, and transitional tensions might have been similarly grounded. They aren't—on the whole, the mainstream fails to recognize the legitimacy of the Indigenous intellectual template and engage with it imaginatively.[6]

Finally, it limits our ability to harness modern Western GIS technology for the benefit of an Indigenous society in transition. The genius of Indigenous technology was the extent to which vast knowledge was embedded in everyday life, congruent with everyday context and resonant with everyday experience. The technology—human memory—relied on constant refreshment through repetition, experience, and social and ceremonial aids. As the physical reality of modern life weakens these methods, alternative technologies can help to maintain the scale and precision of cultural knowledge, which in turn grounds social identity and structures. Where the framework for that cultural knowledge is geographic, GIS technologies provide a uniquely appropriate tool to support documentation, preservation, and propagation.

PRESERVING MARTU KNOWLEDGE— KANYIRNINPA JUKURRPA

Kanyirninpa Jukurrpa (widely known as KJ) is an organization that works closely with Martu communities and since 2009 has been delivering a range of cultural, environmental, and social programs. The largest is the ranger program. The Australian government and other supporters contract KJ to preserve the environmental values of a huge desert landscape. This country now has value to the West, in line with the modern environmental ethos. The Nature Conservancy has identified the desert country over which Martu hold native title as among the most ecologically intact of arid landscapes in the world:

> The arid ecosystems of Australia have the lowest levels of human influence of all arid ecosystems globally and have been identified as one of 24 global wilderness areas and in combination with adjacent Northern savannas, Australia's deserts are recognized as one of the five great wilderness areas remaining in the world, along with the Amazon, Antarctica, Canadian Boreal Forest, and the Sahara Desert. (The Nature Conservancy 2012)

Certainly, Martu country is more intact ecologically than most other parts of the Australian desert.

Preserving Martu country

Modern Western society is willing to invest in the preservation of this rich and diverse ecology. KJ employs over 40 Martu permanently and another 300 annually on a casual basis to work in this country. The ecology has deteriorated since Martu left their traditional life over 50 years ago (Walsh 2008, 186--187). But controlled patch-burning, removal of feral flora and fauna, repairing of water sources, and controlled reintroduction of species are allowing many tracts of the country to revive.

> We have been relocating the rock wallabies here (Pinpi). We want them to breed up around here [figure 7.8]. We have got a radio tracker to find where those rock wallabies have gone [figure 7.9]. They might have shifted around to the other side. We use the cyber-tracker [a GPS-enabled tablet application] as we go looking for pussycat, dingo, and fox; they might eat all the rock wallabies. The satellite will record the location of the dingo tracks for us. The guys coming behind have been throwing out all the baits where those dingoes are. (Samson 2014)

Martu see the ranger program as fulfilling cultural obligations rather than being based in a Western environmental ethic. But the program provides a happy point of mutual value.

> These whitefellas are on our side; they help us Martu men. They know about the permanent waters; they know the waters are giving things to us, to whitefellas and Martu. They [the waterholes] are giving us the game meat, the bush tucker, the kangaroos, and the dingoes. This permanent water holds (*kanyirni*) them and is secretly giving them to us. Wirnpa *Jukurrpa*

Figure 7.8. Carlson Jeffries with a baby rock wallaby.

Photo courtesy of Kanyirninpa Jukurrpa.

Figure 7.9. Shannon Sampson with a radio tracker.

Photo courtesy of Kanyirninpa Jukurrpa.

is doing that. Wirnpa was giving to all of them. It is looking over all of us. (Taylor and Muuki 2009)

Preserving Martu knowledge

The ranger program is complemented by more explicitly cultural programs. The largest of these, the *Kalyuku Ninti* (Knowledge of Waterholes) program takes family groups to their remote ancestral country. Often, this is the first time any of the family has seen these sites and may be the first re-engagement of Martu with particular remote sites since the 1960s. These trips, which can involve up to 60 people, have the character of a pilgrimage. On the trips, the elders provide permission to young people to return and care for the country that has been opened up.

> KJ is helping us Martu. We are bringing our children and family, and we are showing them these places, so they can see them and maybe their children might see these places. We will show them and place them as owners. Our ancestors a long time back used to lead us around these waters here [when we were] children. As a result of that, we know about these waters and dreaming *jukurrpa*. (Taylor and Waka 2009)

The other cultural programs, grouped under the name *Puntura-ya Ninti* (Martu Knowledge), are focused on the collection and preservation of diverse cultural knowledge. Genealogies, which are stored in a specialist genealogy database, map family trees from several generations prior to first contact to the modern day. A long-term waterhole mapping program has located and identified over 1,000 significant sites throughout the desert (figure 7.10).

Information from this program sits in an accessible database that has been loaded onto tablets and is used in the field to inform ranger teams and other KJ staff about places they are visiting or planning to visit. A pilot land use and occupancy mapping program has recorded traditional characteristics and uses of specific sites and precincts. KJ's language program records and preserves language. Information from KJ's history program, which records people's oral histories including stories about travel through the country in *pujiman* days and which also collects archival material about the history of Martu country, is all entered into KJ's cultural and research information database, which was designed specifically for Indigenous people. This database is accessible in each Martu community.

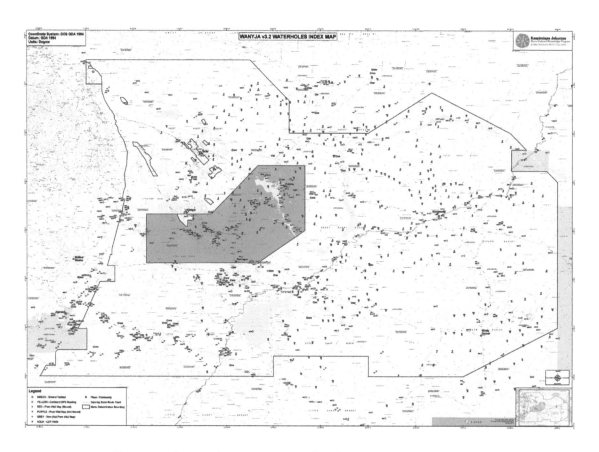

Figure 7.10. A map of water sources within the Martu native title claim based on the *Yintakaja-lampajuya* waterhole map created in 1987.

We got in the chopper and we looked for a waterhole. My grandmother told me to look for an acacia tree growing next to a small creek. We found that acacia tree and saw that water there. We sat down and I took out the recorder. I asked Kumpaya, "What is the name of this place? Whose place is it? Who passed away here?" She told me the name of that waterhole and we recorded that story and we walked around together. I was very happy for my country. Happy to keep the country alive and to keep our language, Many-jilyjarra, alive so it won't get lost. So that I can look after the country, and pass it onto my son and then he can pass it onto his sons.[7] (Girgiba 2014)

All of these programs are integrated. The organizing principle for integration is country: which families belonged to this waterhole, what species occurred here, who traveled here and when, what language does this waterhole speak,[8] what happened here?

These environmental and cultural programs provide a foundation of widespread engagement, which grounds programs more explicitly directed to social development and resilience in the postcolonial context. For example, KJ runs a leadership program focused on adult education. This concentrates on practical information about how the mainstream world organizes and operates, particularly those sectors that have a direct impact on the life of Martu communities. So, for example, fields that for many mainstream audiences would seem arcane form the heart of the program: corporate law, trust law, native title law, government structures and processes, the criminal justice system, financial management, and governance. The aim of this program is to equip young Martu to be confident and adept in negotiating the relationship between their community and the mainstream world. Such a program can only be successful when it is embedded in a culturally resonant suite of activities and development.

COMBINING MODERN AND ANCIENT SPATIAL TECHNOLOGIES

KJ regularly combines traditional knowledge with modern technologies. This marriage of knowledge systems is not ideological—it is the best, the only way, in which the complex work over enormous geographic areas can be effectively planned.

In 1987, Sue Davenport worked with a colleague, Michael Gallagher, and a group of Martu elders to create a map of their country (figure 7.11). On three doors, they traced the sparse visible geographic features of the 50,000 square miles of country: two rivers, a handful of lakes (figure 7.12).

Over several months, the elders sat and sang the country, gradually identifying the locations of over 500 waterholes relative to the rivers, lakes, and other waterholes (figure 7.13). That map, a product of Martu technology, is known as *Yintaka-ja-lampajuya* ("These are our waterholes"). Today, it forms the basis of planning for much of KJ's work. Using vehicles and helicopters, rangers now record the precise locations of those waterholes and record firsthand information from elders about each waterhole, its use, and its significance. Layers of different types of knowledge are added to this map, providing a cross-cultural window on the myriad forms of significance that sites and regions hold.

The KJ programs rely strongly on many forms of technology. Mapping waterholes on country combines searches by helicopter and land vehicles with GPS

Figure 7.11. Sue Davenport draws the Martu *Yintakaja-lampajuya* waterhole map in 1987.

Photo courtesy of Kanyirninpa Jukurrpa.

Figure 7.12. The minister for Aboriginal affairs is shown the Martu *Yintakaja-lampajuya* map by Billy Atkins, 1987.

Photo courtesy of M. Gallagher.

recording, as well as taking elders "flying" using Google Earth. Evidence of threatened and feral species is logged using GPS-equipped tablets. Workplans for rangers on country are prepared using a GIS. The traditional state of Martu country is assessed by combining satellite imagery from the 1950s and 1960s, when Martu still walked and worked their country, with traditional recollections of species location and histories.

Figure 7.13. Waka Taylor points at the old soak Parparnu found during helicopter mapping, 2012.

Photo courtesy of Kanyirninpa Jukurrpa

KJ's manager of fire programs explains the way in which modern technology is combined with traditional knowledge to inform a landscape-scale ecological program of work (figure 7.14):

Rangers have become very familiar with the use of satellite imagery for planning fire (or waru) work on their country. Their intimate knowledge of place is combined with the use of these images to describe the current state of the landscape. This blends contemporary land management and monitoring techniques with Jukurrpa and firsthand site knowledge. (Catt, personal communication, 2016)

This work is very much a two-way exchange, with Martu descriptions of landform and stages of plant regrowth forming the language of conversations, as Martu terminology makes the most sense when talking about their country (despite the difficulties it causes for Western land management staff in pronunciation and recollection).

Younger Martu are proficient in the use of modern technologies such as smartphones and tablets. The enhancement of these devices with mapping applications allows for the caching of Google Earth imagery or the loading of custom maps and is accessible to these young people. These make usable, interactive tools that speed the process of examining and understanding distant and infrequently visited corners of the vast, remote Martu country.

Figure 7.14. Justin Watson uses modern methods for burning, 2015.

Photo courtesy of Kanyirninpa Jukurrpa.

The effect of Martu fire work on the landscape is measured remotely using data gathered in the field. GPS points and track logs give guidance for remote sensing. Staff from the State Department of Parks and Wildlife use satellite imagery to look for fire scars measuring tens of acres in an area that is tens of thousands of square miles. The expertise of external agencies is vital for KJ in building the knowledge and data around the Western Desert. In many ways, this area is the most poorly known to mainstream science in Australia.

Small burn patches, as opposed to vast wildfires, are what fosters ecological diversity in a landscape that, left untended, trends toward a monoculture of spinifex. This network of small patches forms an ever-changing buffer against the large wildfires that ravage the untended areas of country. Traditionally, Martu used fire extensively as a means of actively "farming" the country, belying common perceptions that a hunter-gatherer economy has a relatively passive relationship to the environment (figure 7.15). Over millennia, the landscape adapted to this regular human firing. Making comparisons with aerial imagery taken in 1953 provides a guide on the approximate fire size that would have been a feature of the traditional Martu landscape and gives a guide for what to aim for in the modern day (Catt 2016).

Figure 7.15. Minyawu teaches young rangers traditional firestick burning, 2015.

Photo courtesy of Kanyirninpa Jukurrpa.

Navigation across huge distances of desert country combines GPS, topographic maps, and the memories of Martu elders. Characteristics of sites and landscapes are recorded using photography, video, and voice recordings. The well-being of relocated colonies of threatened species is tracked using satellite tracking devices (figure 7.16). Safety is managed through the tracking of vehicle and team locations through a variety of GPS-enabled devices.

What we are really doing out here, we are learning both ways, whitefella way and Martu way. Learning how to use a GPS, finding waterholes and putting it on the GPS. I think this is really important because the next generation is coming up. For them, when the old people pass away, they will be missing out. So what we are doing now is really good. Doing the same thing mapping country, going back to country, while the old people are still walking. (Robinson 2012)

It's important for me and my kids in the future learning how to use the GPS and mapping in the helicopter. It's really important for me to learn more and more. (Bidu 2012)

Figure 7.16. Waka Taylor shows a group of schoolchildren a *mankarr* (bilby) he has dug up, 2009.

Photo courtesy of Kanyirninpa Jukurrpa.

KJ collects and collates large volumes of disparate data, almost all on the basis of geography or to be linked and organized by geography. Family genealogies are mapped to the country. Oral histories are mapped to the country. Post-contact historical events are cross-referenced to locations. Languages and traditional social groupings are mapped to the country. To the extent allowed by traditional owners, *jukurrpa* stories relating to waterholes and areas are mapped to the country. It simply

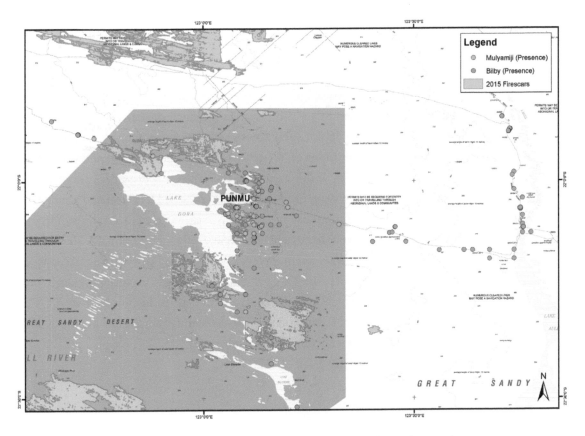

Figure 7.17. A consolidated map shows burning done by KJ rangers and its effect on the prevalence of *mulyamiji* great desert skink and *mankarr* bilby sightings recorded by KJ rangers near the Punmu community.

makes sense to record and organize much of Martu knowledge on a geographic basis (figure 7.17).[9]

Building social resilience

Martu society, like all Australian Aboriginal societies, is resilient. Martu maintain a strong identity, social structure, cosmology, lore, and language. Nevertheless, they find themselves in a situation never previously confronted by their society—a collision with modern European society, technologies, perspectives, and power. In the modern context, we see social resilience as the capacity to forge a path of their choosing to a future that accommodates this collision and yet is positive in their terms. Unquestionably, that requires maintenance of a Martu identity.

The aspirations and sources of inspiration that motivate much of mainstream society do not, on the whole, resonate with young Martu. They have no desire for status in a Western career framework. Instead, their status will derive from their acquisition of cultural knowledge—a different currency. They have no fear of poverty in old age, living within a kinship system in which they are rarely socially isolated. The Martu people do not subscribe to mainstream religion or even altruistic goals shaped by the inspirations of Western society. What they revere—where they find meaning and purpose—is different from the values that drive many mainstream people. The focus of reverence, of meaning, is country and its associated *jukurrpa*.

For an Indigenous community such as the Martu, engaged in a transition from a traditional society and economy to a hybrid society, mixing traditional with Western, resilience is likely to be grounded in their traditional social fabric. They are unavoidably caught in a process of assimilating Western resources and social pressures into their society as a consequence of the confrontation with the Western world. The more the process is one of adaptation and self-determination, the less likely it will be one of dislocation and dysfunction (figure 7.18).

Figure 7.18. Younger men dig out Kiriwirri soak on a return to country trip, under the guidance of Martu elder Waka Taylor, 2013.

Photo courtesy of Kanyirninpa Jukurrpa.

We have seen that, where young Martu find meaning and purpose, they find strength and confidence. As they travel into the country, learning from their elders, they calm and readily state their desire to learn. As their elders entrust them with more knowledge, traditional structures of authority and social reproduction are reinforced. As they are provided with opportunities to ground and develop their personal identities in deeply held cultural values, they develop a sense of role and obligation and a sense of pride, giving them a positive place in their modern world.

> Because of trouble, because of lock up, most boys are going to prison. This job is alright. It will look after them, so they can look after their country. (Booth 2012)

These personal qualities are the bedrock of social resilience. Individuals' resilience—borne of confidence, purpose, and pride—grounds more general social resilience. This is particularly true when the source of individuals' personal development is shared and resonates deeply with the values of the whole community.

The conventional modern Australian response to Indigenous communities' perceived dysfunction is to map out a Western strategy for social development, essentially grounded in economic development. As a rule, this fails. It fails because the assumptions, the underlying mythology, are predominantly Western. A program of pure economic development implicitly appeals to Western values and fears. These do not resonate with young Martu—they have no pull.

This is not to say that a program of economic development would not be beneficial to Martu communities. Rather, for such a program to succeed, it must resonate with the values and mythologies that inspire Martu.

Country as the foundation for community engagement

Our experience with KJ has been that broadscale re-engagement with country has been a catalytic social phenomenon. It has created a culture of participation in communities—regular activity in environmental and cultural activity that is enthusiastically embraced (figure 7.19). It has created the foundations of a new, hybrid economy: engagement with the mainstream economy but through activities of Martu choosing, organized to be congruent with the requirements of Martu society and building on Martu capacity. It has created pride and a striking sense of purpose. It has created a tangible point of connection between the community-based Martu of today and the traditional life of their ancestors—until 2009, very few Martu had visited the remote locations to which their families belonged and which their families left in the second half of the 20th century, whereas by 2015, over 100 young Martu had traveled extensively through these remote landscapes.

Figure 7.19. Martu young people work with Martu elder Kumpaya Girgiba to locate waterholes on a map, 2014.

Photo courtesy of Kanyirninpa Jukurrpa.

The social benefits of Martu re-engagement with country have been independently verified and assessed. Benefits include reduced substance abuse, reduced crime, less incarceration, increased social cohesion, strengthening of traditional authority structures, an increase in community optimism, and a 3:1 social return on investment.

KJ's on-country programs have generated transformative change across the Martu communities. Over the last five years, the programs have produced a wide range of Martu (social and economic) and Jukurrpa (cultural) outcomes. The achievement of these outcomes is entirely dependent on the engagement of Martu on-country. The more time that Martu spend on-country, the greater the value created by KJ's on-country programs (Social Ventures Australia 2014, 3).

When people go out on-country they say, "I'm here, I know who I am and I know where I come from, and I'm going to take charge of my life," and in doing so, they're dealing with the dysfunctional aspects of their lives and their families' lives. So you're dealing with the social issues that are going on in town—but you're dealing with them out on-country—through a social, cultural, and spiritual healing process. (Farmer 2014, 3)

Holding culture

Kanyirninpa Jukurrpa means something like "holding on to culture," or even "preserving and nurturing Martu reality."[10] All of KJ's programs are shaped by and appeal to Martu values, aspirations, and inspiration. At the deepest level, KJ's programs appeal to what Martu hold sacred.

The ranger program is chiefly seen by the West as an environmental program. For young Martu, it is a cultural opportunity. Surveys of young Martu rangers have revealed that their top motivation for participating in the ranger program is to acquire cultural knowledge. The second is to fulfill cultural obligations by looking after the country. The program connects them not simply with the landscapes and sites of their forebears, but with the reproduction of their society, the continuation of a dynamic and sustaining relationship between Martu and their land.[11]

They learn about waterholes, they learn about their families' connections to particular sites and ranges, they learn the mythology and associated ritual, they learn the songs, and they learn about species and their place in particular precincts and sites. Time and again, they speak with reverence about "seeing my grandfather's/grandmother's country." Their work is more based in pilgrimage and learning than in economic advancement (figure 7.20).

Geography is therefore more than simply a template for the storage and propagation of knowledge. Knowledge of and relationship to country are the bedrock of identity. They provide the foundations for a sense of purpose and value.

The preservation of cultural knowledge, embedded precisely in geography, is fundamental to the Martu people's journey in transitioning their society to deal with their modern context. The stability, strength, and resilience of their society through this transition is more likely to be based in the preservation of a strong cultural identity than in any blithe adoption of Western identity.

In the past, this cultural knowledge was stored in individuals' memories. It was transmitted orally. It was acquired over a lifetime of intense personal engagement with a landscape. The acquisition of this knowledge provided individuals' aspirations and sense of status and purpose. The multilayered, interlocking tapestry of

Figure 7.20. Martu, in a respect line, walk to an important spring located about 800 kilometers northeast of Newman.

Photo courtesy of Kanyirninpa Jukurrpa.

stories, song, and ceremony through which *jukurrpa* was engaged with and transmitted was vast and complex. The country provided the intellectual framework for sustaining the scale and richness of this canon of knowledge.

The old ways can no longer sustain the transmission of this knowledge. The modern, more sedentary life of Martu communities and the pressures, distractions, and constraints of their new context compromise the extent to which this cultural wealth can be preserved and passed on using the traditional methods.

What does GIS offer?

New technologies provide new opportunities. They can be appropriated for Martu purposes and their use shaped by Martu priorities. Vehicles provide new means of travel over huge distances. Guns aid in hunting. Recording equipment provides the means of recording and preserving information.

Similarly, geographic information systems provide a means of storing, organizing, and retrieving the complex cultural knowledge that grounds Martu identity and society. They provide a uniquely appropriate technology—layering data on geography, connecting multiple geographic points in different ways, restricting or allowing access to information according to predetermined protocols.

Which families belonged to this waterhole? What language (or languages) does it speak? What stories and songs pass through this waterhole? What happened here in *jukurrpa*? What is the song or story? Where does the songline travel from and to? What ceremonies were performed here? Who is custodian of those ceremonies?

Who are the custodians of this waterhole? In which seasons did people visit here? What species were abundant? Where did they camp? Which parts were restricted to initiated men or to women? Where did people travel here from? Where did they travel to? What happened here?

To the extent that this knowledge can be preserved, over vast tracts of country and for over 1,000 waterholes and other sites, Martu society can remain richly connected to its past. The alternative is disconnection from the past—dislocation from country and culture.

GIS does not provide a modern substitute for the stories and songs through which *jukurrpa* and domestic knowledge have always been transmitted. But the breadth and depth of knowledge of those stories and songs, and the intense depth of knowledge of country, is difficult to sustain when Martu are no longer living and walking every day through the landscapes the songs describe. GIS provides a modern adjunct to traditional methods. GIS can incorporate detailed information about sites, precincts, and vast areas. Geographic information systems can link locations to stories and songs as well as to rich and diverse data. They can combine and layer different forms of apparently unconnected information. They provide a persistent aid to memory, where memory can no longer be relied on as the sole repository.

Figure 7.21. Martu Leadership Program participants pose in front of the Australian Parliament House after meeting politicians.

Photo courtesy of Kanyirninpa Jukurrpa.

This is an imperfect solution to the potential (and actual) loss of cultural knowledge. But it is a solution that provides communities with the means to preserve much more than would otherwise be the case. It allows that preservation to retain the primacy of country as the organizing principle for knowledge. It emphasizes the breadth and richness of the mapping of *jukurrpa* to geography.

GIS provides a unique technology to support the preservation and use of cultural knowledge. It therefore provides a unique tool to help sustain and build social resilience in Martu communities, particularly if appropriated by Martu people, with its use designed by them. It is uniquely appropriate, because it shares the use of country as a reference point, a framework for organizing knowledge. That focus links the ancient and the modern. It can help ancient cultures to sustain their distinctive identities and relationship to country (figure 7.21). In that, they find strength.

CONCLUSION

This chapter, and much of KJ's work, are grounded in the following perceptions, built on working with Martu for almost 30 years.

The social resilience of a community is substantially built on individual people's sense of their social identity and role. If these fracture, the community decays. Identity and role are grounded in culture. They are defined within culturally set parameters. Culture, in turn, requires the maintenance and valuing of cultural knowledge. So the vitality of a people's culture has a direct relationship to the social resilience of their community.

Martu people are living through a collision of cultures. Their traditional culture and knowledge were constantly reinforced by their manner of living—in a deep and intense daily relationship with their environment. Their lifestyle is now a hybrid, retaining very strong and deep elements of traditional culture but within a modern, more sedentary context that can undermine the long-term retention of cultural strength and vitality.

Martu are not finding strong identity and roles purely within the mainstream world—it is not their world. The resilience of their communities is manifestly strengthened by the current efflorescence of cultural knowledge, which has been enabled through a physical reengagement over the last five years with their vast, remote, and rich landscapes. Those landscapes are the templates of cultural knowledge; they hold Martu *jukurrpa*—cosmology, lore, families' and individuals' sense of belonging, language, sacred places, and sacred beings.

In their modern, hybrid context, modern technologies help to retain the richness of Martu culture: photographs, film, sound recordings, and documents. These create a hybrid technological foundation that Martu people have adopted, marrying their traditional intellectual methods and content with modern, Western devices.

GIS technology is one such device. It is uniquely matched to Martu knowledge frameworks, helping people to record, organize, and deliver a range of landscape-based knowledge. In Western terms, the landscape is the template. In Martu terms, the landscape literally holds the knowledge—it is not simply an intellectual framework but a living manifestation of knowledge and truth. Either way, the landscape holds the key.

Martu are not unique in this. All Aboriginal groups in Australia have a similar deep-rooted relationship to and identity with their country. Many other Indigenous peoples around the world have such a relationship. That relationship is profound and should be valued inherently, as part of humanity's richness.

But Indigenous relationship to country can also be valued by non-Indigenous societies for simple, practical reasons. Knowledge of country supports culture, culture supports society, and strong society provides social resilience.

Tools, such as GIS technologies in the case of Martu, provide Indigenous societies and people with the means to nurture what makes them strong. These tools allow them to maintain what gives them identity despite postcolonial trauma and dislocation. This technology should be harnessed to help those societies retain their strength and resilience.

REFERENCES

Bennett, Willie. 1987. Personal communication.

Bidu, Jessica. 2012. *Kanyirninpa Ngurrara 2012*, DVD produced by Kanyirninpa Jukurrpa. www.kj.org.au.

Booth, Ned. 2012. *Kanyirninpa Ngurrara 2012*, DVD produced by Kanyirninpa Jukurrpa. www.kj.org.au.

Cane, Scott. 2013. *First Footprints*. Sydney: Allen & Unwin.

Catt, Gareth. 2016. Personal communication.

Davenport, Sue, and Peter Johnson with Yuwali. 2005. *Cleared Out*. Canberra: Aboriginal Studies Press.

Farmer, Darren. 2014. Quoted in *Social Ventures Australia* 2014, 3.

Girgiba, Clifton. 2014. *KJ*, DVD produced by Kanyirninpa Jukurrpa. www.kj.org.au.

Gould, Richard A. 1969. "Subsistence Behaviour Among the Western Desert Aborigines of Australia." *Oceania* 39 (4): 253–274.

James, Diana. 2005. *Kinship with Country*. Unpublished thesis, Australian National University.

James on behalf of the Martu People v Western Australia. 2002. FCA 1208. September 27, 2002.

Myers, Fred R. 1986. *Pintupi Country, Pintupi Self.* Washington, DC: Smithsonian Institution Press.

Robinson, Lindsay. 2012. *Kanyirninpa Ngurrara 2012*, DVD produced by Kanyirninpa Jukurrpa. www.kj.org.au.

Samson, Errol. 2014. *KJ 2014*, DVD produced by Kanyirninpa Jukurrpa. www.kj.org.au.

Social Ventures Australia. 2014. "Social, Economic and Cultural Impact of Kanyirninpa Jukurrpa's On-Country Programs." Melbourne: Social Ventures Australia. http://socialventures.com.au/assets/2014-KJ-SROI-Report-FINAL.pdf.

Taylor, Muuki. 2009. Personal communication.

Taylor, Waka. 2009. Personal communication.

The Nature Conservancy. 2012. *Martu Country: Healthy Country: Assessment of Conservation Values*, unpublished report.

Tonkinson, Robert. 1991. *The Mardu Aborigines*. New York, NY: Holt, Rinehart & Winston Inc.

Veth, Peter. 1989. "The Archaeological Resource of the Karlamilyi (Rudall River) Region." *The Significance of the Karlamilyi Region to the Martujarra of the Western Desert*, edited by G. Wright. Perth: Department of Conservation and Land Management.

Walsh, Fiona. 2008. *To Hunt and To Hold*. Unpublished thesis, University of Western Australia.

Williams, Rosie. 2009. Personal communication.

NOTES

1   Our view of resilience in this context is discussed in a later section.

2   Increasingly, that is the way that Martu see themselves. Where once they may have identified by language group, now they tend to identify with the political, social, and economic bloc represented by Martu.

3   In some orthographies, this is also spelled *tjukurrpa*.

4   Myers was writing about the neighboring Pintupi people, whose language and society are very close to that of Manyjilyjarra, the easternmost of Martu.

5   Two judges in the High Court's major native title decision now known simply as
     Mabo spoke of ". . . the conflagration of oppression and conflict which was, over the
     following century, to spread across the continent to dispossess, degrade and devastate
     the Aboriginal peoples and leave a national legacy of unutterable shame." Deane and
     Gaudron JJ, *Mabo v. Queensland* (No 2) (1992) 175 CLR 1 (3 June 1992) at para 50. In
     the 2016 annual report on the progress of Indigenous advancement, the prime minister
     of Australia noted: "We pride ourselves on having built an egalitarian country where
     everyone has the same chance to realise their dreams and to fulfil their potential. But it
     is not until Aboriginal and Torres Strait Islander people have the same opportunities
     for health, education and employment that we can truly say we are a country of equal
     opportunity." Commonwealth of Australia, Department of the Prime Minister and
     Cabinet, "Closing the Gap," Prime Minister's Report 2016 at 3.

6   One exception to this is a ranger program, funded by the Australian Government,
     under which Indigenous rangers are employed to provide environmental services on
     the traditional country. However, this program only funds about 800 ranger positions
     over the entire continent, and the roots and reasons for its widely acknowledged
     success have not been seen as an indicator of how other, disparate programs might
     similarly succeed.

7   A film about this subject, which includes this quote and information about helicopter
     mapping, may be found at www.kj.org.au.

8   In the Martu world, language is not a property of people but of place. Language is
     a characteristic of country, embedded in the *jukurrpa*. Many people identify with
     and speak two or more dialects, because of affiliations with different country. So a
     person may say "My father was a Mangala/Manyjilyjarra man and my mother was
     Manyjilyjarra/Warnman."

9   Databases are maintained by KJ and distributed to Martu in each community via
     publicly accessible computers.

10   "Kanyirninpa Jukurrpa" is not suited to easy translation—neither word has a true
     equivalent in English.

11   For the full Social Return on Investment Report, see http://socialventures.com.au
     /assets/2014-KJ-SROI-Report-FINAL.pdf.

→ CHAPTER 8

Developing resiliency through place-based activities in Canada

KEVIN O'CONNOR AND BOB SHARP

INTRODUCTION

This chapter focuses on how youth develop resiliency through their involvement in an array of place-based community projects. Within these projects, spatial features of place, using location, mapping technologies, and geographic information systems (GIS), play a significant role in understanding the problems and complexity of challenges faced by communities. "In a globalized age where notions of interdependence, interconnectedness, and common destinies abound, the 'local,' with its diversity of cultures, languages, histories, and geographies, continues to exercise a powerful grip on the human imagination. The ties that bind us have global connections but are anchored in a strong sense of locality or community" (McInerney 2011, 3).

We first describe the educational setting permitting students to take part in such a range of activities outside school settings. Two case studies are presented showing how GPS, mapping, and GIS become an integral part of the science programs and how they are used in place-based and experiential educational activities. This educational process develops skills and knowledge, creates challenge, encourages persistence, and fosters relationships within the class and with community members, thereby contributing to the creation of resiliency in the youth. The integrated studies approach illustrates how involving students in real, complex community questions focused around location and culture can be a success story as local people look to define their own community challenges. The success of local involvement emanates from local youth and their families' ongoing efforts, flexibility, and resiliency in trying to create solutions to local environmental problems and issues.

The high school science program involved in these place-based activities integrates chemistry, biology, geography, art, and environmental monitoring taught through a series of in-school classes and a range of field studies related to place, community, cultural, ecological, and traditional knowledge. In such a learning environment, students develop a recognition of the relationship of self, community, and the natural world (Kincheloe 2005). The relationship to social and spatial place in the context of real local issues provides students with community engagement and the capacity to apply learning from one context to another. This resiliency finds a genesis in community-based inquiry.

Defining resilience

The conditions that give rise to responsible environmental and social behaviors require embracing the principles of resilience. These are a major focus of place-based educational initiatives (Louv 2005; O'Connor 2009; Sobel 2004). Through place-based education work, we have found that creating learning environments that embody challenge, teamwork, learning from mistakes, development of new skills, and significance develops resilience and this, in turn, develops the individuals' and communities' capacity to address complex persisting concerns.

Since there are many definitions of resilience, we choose to use the Stockholm Resilience Centre definition: "Resilience is the capacity of a system, be it an individual, a forest, a city or an economy, to deal with change and continue to develop. It is about how humans and nature can use shocks and disturbances … to spur renewal and innovative thinking. Resilience starts from the belief that humans and nature are strongly coupled to the point that they should be conceived as one social-ecological system" (Stockholm Resilience Centre 2017). Social conditions that give rise to a focus on citizenship suggest that critical thinking is a foundational skill along with resilience. Pedagogical foundations associated with place-based initiatives integrate these conditions and are foundational for school instructional strategies (O'Connor and Sharp 2013).

PLACE-BASED EDUCATION

Place-based education (PBE) provides the learning context. PBE is an approach to teaching grounded in the context of community, both natural and social. It connects place to self and community. The field has emerged from the strong roots laid by 30 years of environmental education in North America (Orr 1994; Raffan 1993). PBE provides a purpose to the knowledge and reasoning taught in schools, thus providing a contextual framework for much of the curriculum (in other words, it gives meaning to the studies) and engaging students in the conditions of their own reality. R. W. Tyler examines an educator's ability to influence the environment to promote learning: "It is desirable that the problems be set up in the kind of environment in which such problems usually arise in life. This is more likely to result in [the student] viewing this as a real problem worthy of his effort to solve" (Tyler 1949, 69).

The two case studies, selected from more than 180 place-based activities, illustrate how place and real community problems contribute to the development of resilience. These PBE initiatives use critical knowledge of social, environmental, and political issues and associated action strategies, increased locus of control, sharing community attitudes, trial and error evaluation processes, and an individual's sense of responsibility within a community as strategies that develop resilience. The place-based activities draw on the local understanding and knowledge of a particular place and its resources. Collectively, these approaches teach ways to address complex social, environmental, and ecological problems. The strategies are developed and practiced in the context of community places and community problems. Such repeated pedagogies leave the student with a deeply felt attachment to such processes. Our research illustrates how such processes are the fundamental strategies currently applied by our former high school students, some as long as 25 years ago, for addressing complex problems. An experiential education, getting out into the field and the community, can foster responsible and resilient citizenship. Involving place-based learning activities encourages trial and error processes in which data collection, reflection, and action are important antecedents to responsible and resilient citizenship.

A crucial condition to critical pedagogy is the need for a context to be relevant and therefore sustainable (Gruenewald 2003; Penetito 2009). Community issues that frame place-based learning provide the context for critical thinking, situational conditions, and attributes such as locus of control. Place-based educational activities focus on environmental and social values, situational characteristics, and

psychological variables, as community action is open to a range of varying and com-
peting interests (Barr 2003). This speaks to experiential and place-based science programming, strategic inclusion of adversity, and resilience into curricular delivery
and content.

A survey of the literature on PBE reveals characteristic patterns to this still-evolv-
ing approach that make it distinctive. It emerges from the particular attributes of a
place and how problems related to place are often complex and layered. The content is specific to the geography, ecology, sociology, politics, and other dynamics of
that place (Gruenewald 2003). This fundamental characteristic establishes the foundation of the concept and it is in this context that GIS exhibits particular value in
examining and analyzing such problems.

Place-based learning with GIS

Place-based learning with GIS is inherently multidisciplinary. Such learning is also
inherently experiential. In many programs, this includes a participatory action or
service-learning component; in fact, some advocates insist that action must be a
component if ecological and cultural sustainability are to result. It is reflective of an
educational philosophy that is broader than "learn to earn." Economics of place can
be an area of study as a curriculum explores local industry and sustainability; however, all curricula and programs are designed for broader objectives.

Such place-based learning connects place with self and community. Because of
the ecological lens through which place-based curricula are envisioned, these connections are pervasive. These curricula include multigenerational and multicultural
dimensions as they interface with community resources (O'Connor 2009).

Cohort of students versus subjects of students

Most conventional high schools in North America offer an array of courses taught
by a teacher specialist who may see more than 100 students a day, and those students
are often with different groups of students from one period to the next. This pattern
of school organization makes integration across subjects and collaboration among
students across subjects problematic. Developing these relationships in schools can
be difficult where organizational structures minimize opportunities for extended
personalized relationships, as is often the case in "factory-model" schools designed
a century ago for efficient batch processing of masses of students (Tyack 1974).
The design of most United States secondary schools is particularly at odds with the

needs of adolescents, as high schools deemphasize personal connections with adults and engage in intense evaluation and competitive ranking of students (for example, in academic tracking or tryouts for clubs and activities) just as young people are most sensitive to social comparisons and most need to develop a strong sense of belonging, connection, and personal identity (Eccles and Roeser 2009). Unless mediated by strong relationships and support systems, these conditions interfere with learning, undermine relationships, and impede opportunities for youth to develop skills to succeed (Osher and Kendziora 2010). This chapter refers to this modeling of schooling as subject schooling and it carries both the traditions and problems identified above.

Ecological changes that create personalized environments with opportunities for stronger relationships among adults and students can create more productive contexts for learning (Felner et al. 2007). For example, small schools or small learning communities with personalizing structures—such as advisory systems, teaching teams that share students, or looping with the same teachers over multiple years—have been found to improve student achievement, attachment, attendance, attitudes toward school, behavior, motivation, and graduation rates (Bloom and Unterman 2014; Darling-Hammond, Ross, and Milliken 2006; Felner et al. 2007). The cohort model—in which students participate in an integrated studies approach to a suite of courses, work together in many field studies, and are freed from the school bells—represents such a personalized environment. The integrated studies programs, organized around themes that appeal to students, can make even large high schools into smaller, more intimate settings.

Experiential and place-based learning

There are a variety of educational approaches that involve integration of curriculum across disciplines, addressing local situations as a means of providing relevance to school studies, and using collaborative approaches to schooling. Experiential and placed-based approaches have been a logical extension to the rural educational setting (O'Connor 2009). Experiential and place-based initiatives have become a major factor in education, as many Indigenous and rural northern Canadian communities move toward greater autonomy and self-determination: to encourage students to be aware of and feel responsible for the lands their ancestors have occupied and to better prepare and encourage students for employment opportunities that exist within rural territories and beyond. Initiating experiential and place-based approaches requires changes in the ways in which schools organize their time and instructional processes. These approaches require flexibility in scheduling integrating and combining courses and changing relationships between student and teacher.

Experiential and placed-based science initiatives were created to address the lack of success and disengagement among Indigenous and rural students in formal schooling and have expanded to take the place of much of the conventional schooling. Such initiatives take many forms and themes. Within the sciences, a holistic form of science education has evolved: a form that values the importance of place and the role of cultural knowledge within the disciplines of what some would consider formal science and also more cultural interpretations such as traditional ecological knowledge. This chapter examines the long-term effects found in multidisciplinary approaches to science and restructured school experiences to build on the strengths of such communities.

Experiential Science Grade 11 program

The Experiential Science Grade 11 (ES11) program was created as a pilot in 1994 and is still in existence as a territorial education model. The ES11 program is a Yukon territory public school program of studies open to Grade 11 students. Rural and Indigenous students from a wide range of schools choose to take part in the program. Students spend 35 to 45 days of a 93-day semester conducting field studies related to community issues and strategically connected to integrated curriculum science-related studies. During a 10-year study period, 357 students have participated in the program and three educators have taught the ES11 program. About 200 place-based activities covering a wide variety of issues have been the focus of student learning over the decade of research.

The ES11 program integrates Biology 11 (a survey course including studies related to population ecology), Geography 12 (studies of atmospheric dynamics, geomorphology, and resource utilization), Chemistry 11 (introduction to quantitative chemistry), Art 11 (focused on scientific illustration and landscape), Field Methods 11 (applied studies in environmental monitoring protocols), and Physical Education 11 (focused on physical well-being and outdoor education). Students are engaged in two full-day labs a week (chemistry and biology), housed in the local postsecondary institution, Yukon College. Field studies expose students to a wide variety of experts associated with a range of resource management issues. Rigorous field methods, reliable well-kept data, and valid scientific methodology are the foundation of the program. Through the use of environmental science-related activities and an integration of traditional ecological knowledge, students develop an enlightened recognition of the proper relationship of self, community, and the global world. Students collect field data and analyze various aspects of environmental study issues before developing strategies to address and take action related to community concerns.

During an ES11 semester, students participate in a wide variety of place-based activities, often in the company of scientists who have been working in a related field. They take part in an intensive month-long trip that involves field studies and community activities conducted in a range of settings. Most of the activities involve environmental monitoring and most are longitudinal in nature, as they span over a period of years. The community issues students address during their time in the ES11 program are typically characterized as PBE initiatives. The ability to infuse an outdoor activity with related environmental field studies benefits the whole educational enterprise. The linking of environmental field studies with an outdoor pursuit gives both the study and the activity additional educational value and meaning. In addition, field studies reinforce both labs and lectures in specific subjects, addressing a traditional education problem: integrating theory and practice (Dewey 1938).

Courses such as geography, survey biology, quantitative chemistry, ecology, and environmental studies are often integrated and lend themselves to field studies that link to a range of outdoor activities. The field studies approach often takes on the mantel of PBE since many of the field studies are centered on responding to community concerns, studying and collecting data, and proposing possible responses to the community-defined problem. Addressing real topics and finding ways to apply the prescribed learning outcomes to these studies have proven to engage students in ways that secure knowledge, develop resiliency, and strengthen positive community attitudes. In this respect, including field studies with outdoor pursuits has been proven to be a successful educational approach (O'Connor 2010; Raffan 1993; Woodhouse and Knapp 2000).

PLACE-BASED EDUCATIONAL ES11 PROJECTS

This section illustrates a focus on science projects that both involve and impact the local communities. The common link between both case studies is that they address issues in the local physical environment and use spatial technology in the form of GIS as part of the research process. Students engage in hands-on learning by conducting field research in communities that they engage with as part of the research process. Additionally, multiple research methods relying on physical science and social science are used across the case studies to provide a complete view of the issues examined.

Case study 1—Ibex River Salmon

The Ibex River Salmon Enhancement Project was undertaken by Experiential Science 11 in partnership with another school and various community groups that had undertaken a salmon restoration program on the Ibex River between 1998 and 2004. The Department of Fisheries and Oceans (DFO) provided a Habitat Restoration and Salmon Enhancement Program (HRSEP) grant for the work that was to be undertaken. The following outlines our restoration endeavors.

The Ibex River (figure 8.1) is a remote river only accessible by four-wheel-drive vehicle in the summer and snowmobile in the winter. The Ibex River is a small river located about 30 kilometers west of Whitehorse, Yukon territory, Canada. The river has undergone a number of changes over the past 40 years, which have impacted salmon populations. Prior to that time, the Ibex River had been a valued food source for the *Kwanlin Dun* First Nation and the Champagne and Aishihik First Nations. The redirecting Fish Lake stream, the fire of 1958 followed by a regeneration aspen stand, and declining beaver pelt prices resulted in a lowering of Ibex water levels and an explosion in the beaver populations. Increased access to the river by the Ibex tote road also appears to have given rise to an increased salmon take. The combination of these factors has resulted in a decrease in the salmon population on the Ibex. The intent of this restoration program is to re-establish traditional salmon populations along the Ibex River. Our long-term objective was to create a restoration program that would see salmon stocks rebound to historical levels.

In the first year, students conducted a GPS mapping exercise of the Upper Ibex River system down to the confluence with the Arkel River, identifying some of the more obvious obstructions to salmon upstream movement. Our plan was to develop in-stream incubation facilities that could be used in remote settings. During the first years, we proposed to use the stream-side incubation facilities in Whitehorse to raise eggs from the Takhini River to fry stage as a first step in restoring the salmon populations on the Ibex River. The initial inventory of spawning salmon along the Ibex proved to be a challenge. Some of the beaver dams were more than 10 feet high. We mapped access points to various reaches of the river system. We also identified reaches in the river that may be characterized as potentially high-quality spawning

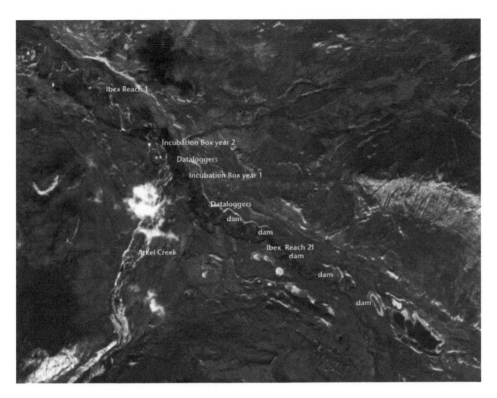

Figure 8.1. Ibex Valley.

GeoYukon, with plot locations indicated by R. Sharp.

areas. We also embedded four-channel data loggers so that substrata temperatures could be determined in areas considered to be best suited for an in-stream incubation facility. During this survey, we identified logjams that are obstructions, redds (salmon nests where the female lays her eggs), the salmon-created spawning beds, and spawning or dead salmon. It was our intention to cut paths through selected barriers to reduce obstruction to salmon movement and increase access to the regions of the Ibex being surveyed. It was also our intention to set "G" traps in a variety of locations during the winter months to assess fry populations on the Ibex. We may have missed redds or salmon in our survey. The river is quite difficult to paddle and much of our attention during our surveys was spent paddling, avoiding obstructions, and keeping afloat.

By the end of the first year, we had canoed and surveyed 40 kilometers of the river on two occasions with visits to selected segments a number of times. We have developed enriched 1:25000 digital maps showing dams and obstructions, locations of the data loggers, and placement of the incubation boxes.

DFO encouraged us to make and try an incubator box before placing salmon eggs into the box. Following ES11 visits to the Campbell River Quinsam Hatchery, where they were experimenting with self-release in-stream incubation boxes, we decided to explore this as an option with required modifications. We recognized that we would need to explore sites for appropriate substrate conditions for over-winter survival, and we would need to design an incubator that would not freeze in the winter and would not be removed by the spring breakup ice flow.

ES11 students in 2001 designed and installed an incubator made from a 45-gallon plastic food barrel, cut to 50 centimeters in height. ES11 classes conducted seasonal monitoring of the site. The first barrel was removed by spring ice, so the second was embedded deeper and moved to a quieter stream location. Access to the Ibex location was difficult. It is about a one-hour drive in a four-wheel-drive vehicle and about a one-hour walk along a route without an established trail. We also surveyed (with students on bikes with GPS) about 25 kilometers of the Ibex Valley, identifying access roads to the river.

We installed two data loggers in the Ibex River, one above and one below the confluence of the Arkel River, to monitor the in-stream temperatures at the location of the incubation box in conjunction with DFO personnel. ES11 students developed two studies with Yukon Energy "Green Power" to explore the use of solar panels and an in-stream generator as a means of providing heat to the boxes, if needed. The Onset four-channel data loggers were placed in the substrate, one in the air and the other in the water. These were placed in the Ibex in November 1999 and picked up in May 2000. We logged temperatures in the stream to see whether there was a warming groundwater influence in the Ibex because of the flow from the Arkel River. Data was collected over a seven-month period. The data shows a period when stream temperatures dropped to 0 degrees Celsius for about 50 days.

In 2017, DFO, in partnership with *Kwanlin Dun* First Nation and the Champagne and Aishihik First Nations, reinitiated the salmon restoration program following many of the processes the ES11 classes had developed.

Case study 2—Cowley Lake

The second case study focuses on Cowley Lake in the Yukon. For this case study, the project began as a response to a community request and, over time, morphed into many different educational field studies. In 2004, the Hamlet of Mount Lorne Local Advisory Council approached the ES11 class with a local concern and a request for the class to undertake a series of studies that would help the community understand what was happening to Cowley Lake. They asked, "Why has our lake level dropped more than a meter over the past five years?" The community had asked government agencies to look into their concerns but received little support. Instead of giving up on these concerns, the community turned to the ES11 class, asking them to study the issue. These actions show community resilience in trying to find answers to their questions. It also demonstrated the reputation that the ES11 class and the Wood Street School's programming had developed within the Yukon territory and beyond. The ES11 class decided to try to answer the community's big question. The class began with discussions with community members and turned to discussions with geologists, ecologists, and historians. Approaches to this kind of community question require flexibility, trial and error approaches, innovative thinking, and persistence. In the case of the Cowley Lake study, we began with one set of studies, but over time these became a number of related studies. In this process, the expanding of the community questions demonstrates resilience.

ES11 classes developed a study plan to better understand the lake. This plan included three major parts: (1) an understanding of the history of the lake since the last major glaciation, (2) an analysis of the lake water balance, and (3) studies related to water quality. Studies related to the history of Cowley Lake involved an examination of the oral history, an analysis of water balance, and a focus on studies related to water quality.

Cowley Lake is a shallow horseshoe-shaped lake located about 30 kilometers south of Whitehorse, Yukon territory, Canada. Between 1998 and 2004, the lake level had dropped more than a meter. A number of human activities appeared to have contributed to altering lake levels, but none over this time span have removed substantial amounts of water from the lake. The ES11 class proposed to undertake a more detailed long-term analysis to understand processes that have contributed to changes in lake levels. Such analysis included the following: (1) studying the drainage basin of the lake, (2) conducting analysis of intake and outflow, (3) studying the long-term history of lake levels through study of the paleolimnology, (4) conducting historical analysis focused on how human actions may have altered the lake over the past 100 years, (5) determining flows and dynamics through interviews and review of the historical record, and (6) exploring biotic characteristics of the lake and how these may have changed over time.

This project represented a collaboration between ES11, the Cowley Lake neighborhood, the Mount Lorne Hamlet Council community, University of Alberta, and Yukon College. The university participant provided training and extension education to groups of northern high school students who were, in turn, addressing issues of community concern. Project funding came from the International Polar Year and has involved about 400 students over the course of the studies.

Students who participated in the project used a variety of research methods. These included discussions with community members and a review of records related to human impacts on the lake over the past 120 years. Following that, paleolimnology studies provided insights into past environments and historical lake levels. Studies related to the water balance of Cowley Lake involved four aspects: (1) mapping and measuring the inflows and outflows of Cowley Lake, (2) conducting a bathymetric mapping of the lake, (3) obtaining systematic measurements of lake levels on both sides of the rail line, and (4) mapping possible changes to the inflow and outflow streams.

Studies related to the water quality of Cowley Lake involved four-season sampling related to water chemistry, including pH, dissolved oxygen, conductivity and turbidity, and summer and winter analysis of aquatic invertebrate populations. Many of these studies involved the use of GPS to identify specific locations and to map many of the field studies.

ES11 students conducted interviews with a geologist who had studied the formation of the region. The formation of Cowley Lake dates back to the deglaciation of the Yukon Southern Lakes approximately 11,000 years ago. During the period of glacial retreat, the melt from glaciers occupying Bennett Lake and the Wheaton and Watson valleys drained north through the broad valley that is now a chain of lakes. Cowley Lake is the central lake in this chain and represents a meander in this outflow river. The image below shows the course of this melt river during this period. The melting of the glaciers filled Tagish and Marsh lakes, allowing the water to flow through the path currently followed by the Yukon River, leaving only traces of the river that flowed through the Cowley Valley (figure 8.2). Studies in paleolimnology shed light on these dates. Once the lake developed, diatom populations expanded. By examining this transition period through studying lake sediment cores, we are able to get more specific dates on these events through carbon-14 dating of organic materials deposited at this time.

Interviews with Yukon First Nation Elders reported that salmon and trout were to be found in Cowley Lake before the construction of the rail line. In 1898, the White Pass and Yukon Railway (WPYR) was built to access the Klondike. The construction of the WPYR appears to have redirected the inflow and outflow streams around Cowley Lake. The rail bed bisected two ends of the horseshoe-shaped lake, changing the flows into and out of the lake. If culverts were installed at this time, they would have been made from wood. If such culverts were in place, they have long since collapsed and filled in. The class mapped the paths of Dugdale Creek and Cowley Creek and monitored the flows in these creeks. They also monitored precipitation in drainage basins using snowfall studies in three locations, with snow depth and water equivalence measurements at established sites at different elevations within the basin.

Between 2008 and 2011, beavers dammed Dugdale Creek, backing water up so that it overflowed the rail track, and raising the level of Cowley Lake so that by 2010, levels had risen to historical levels. In 2012, some residents expressed concern that the lake levels may exceed the historical highs, so they asked the Yukon Government Game Branch to remove the beavers from the section of Dugdale Creek that had caused the flooding. In 2007, the Hamlet of Mount Lorne Local Advisory Council (LAC) requested the WPYR to play a role in these studies and in possible remediation efforts, since the track traverses two sections of the lake.

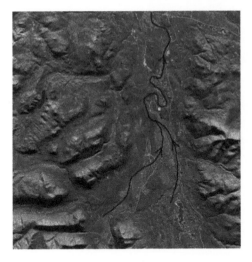

Figure 8.2. Deglaciation drainage pattern.

GeoYukon, with channel overlay by R. Sharp.

At the behest of the LAC, the community requested access to White Pass' files relating to the original construction and subsequent upgrades to the rail bed where it transects the ends of Cowley Lake. In addition, WPYR provided financial contributions to assist in the ongoing needs to purchase scientific data collection and analysis equipment and help defer commercial laboratory expenses, which was greatly appreciated. As part of the studies, ES11 students interviewed a biologist with a home overlooking the lake. He had recorded more than 20 years of observations on spring bird arrival dates around Cowley Lake. He shared these records with the students and these, in turn, were posted on the Cowley Lake website.

A study to determine the depth of the lake involved mapping and GPS use. Bathymetric charts usually show lake bottom relief as depth contours (figure 8.3). The mapping of the Cowley Lake bottom involved nine canoes, GPS, and weighted lines. The canoes spread out from the shore to approximately the middle of the lake. The canoes moved in concert around the lake, taking depth and location measurements about every 100 meters. The waypoints were then converted to the depth recorded, and the points were plotted on a georeferenced image of Cowley Lake. Approximately 1,000 depth points were plotted using this method. The results revealed a mean depth of 2 meters.

Figure 8.3. Cowley Lake bathymetric survey.

GeoYukon basemap with plots overlaid by R. Sharp.

ES11 conducted core sampling of lake sediments at a number of locations to determine a long-term history of lake levels through paleolimnology. These cores were taken from shallow parts of the lake to see if there were layers of terrestrial grasses through the sediment column. This would indicate periods when lake levels had dropped to lower than the depth at which they were sampled, but no such layers were found. ES11 also examined the environmental history of the lake through analysis of diatom populations found at different depths in the core samples. Students were given instruction in the collection and analysis of diatom populations by a University of Alberta professor. The analysis of cores has been an ongoing activity over the past seven years. This, in turn, led to discussions with the Yukon territory archaeologist who was interested to know when the lake stopped being a river and became a lake. The archaeologist encouraged the ES11 class to identify the transition samples and collect organic material that could be carbon dated from this transition period.

Currently, Cowley Lake studies carry on, building on the work of former students. An understanding of spatial and social contexts of place are central to this study. Three groups of ES11 students have made presentations to community members and to residents who live along Cowley Lake. These case studies speak to how students value their understanding of this place from both a spatial and societal perspective and how these educational studies actively contribute to community understanding of place and space. Both GPS and GIS have contributed to the understanding of place. Many features of the students' studies have remained with them as general principles that have been applied to other situations. This capacity to reapply principles they have internalized describes the resilience educators seek and ultimately which rural community members can benefit from an increased participation in the research project and sharing of data and results back with the community.

IMPACT OF ES11 ON POSTSECONDARY PATHWAYS
Research context

We describe two related longitudinal research projects (O'Connor and Sharp 2013) that examine the extent to which an experiential education initiative has fostered responsible and resilient citizenship through place-based activities. The research project follows students from 10 to 20 years following their participation in an integrated secondary science program focusing on place-based activities in which science data collection and peer-driven social, political, and environmental actions are promoted.

The initial stage of research followed 357 students through high school and 150 students in their postsecondary ventures. This does not represent the proportion of students carrying on to postsecondary education, but rather identified gaps in the records of postsecondary grants program. The Yukon provides students with up to 170 weeks of postsecondary grants over their lifetime, provided they continue to meet eligibility requirements. Information following the 170 weeks of the grant program has been obtained through interviews of students and parents.

Methods for the educational study

This longitudinal mixed-methods research (Creswell and Clark 2011) examines the extent to which various educational strategies foster long-term commitments to responsible citizenship. The research followed two distinct yet related paths. The first was a tracking of the postsecondary educational activities of the rural and Indigenous students who took part in the ES11 program. We were able to compile information related to students' choices of postsecondary institutions, program of studies, course results, and employment. Here, we discuss the preliminary results of an initial survey, which combined with interviews and anecdotal discussions with many of the former students, pointed to the value of early student involvement in place-based activities, leading to social and environmental action and responsibility. We conducted "a spiral of cycles" (Lewin 1947) in our research, thus the preliminary results allowed us to refine not only the questions but also the research techniques for subsequent surveys and interviews. This gave rise to the second stream of research designed to collect more detailed information about the choices and actions of this cohort of students and how their ES11 experience may have influenced their subsequent values and actions. To this end, an intensive survey addressing many topics, including questions related to their values and actions associated to active citizenship, was distributed to students of this cohort. At present, both these research projects are ongoing. This chapter reports on our preliminary findings.

Many features of this research are possible because of the size and nature of the community. The city of Whitehorse, in the Yukon territory, Canada, is a relatively small, isolated northern city. Many students opt to continue their postsecondary education in different "southern" universities and technology institutions. The Yukon Grant program provides a 170-week tracking of postsecondary education. Students are required to maintain and report their academic standing to qualify for the ongoing grants. Our research has been able to use the grant program, persistent friendships, and resident parents to assist in our data collection and to keep in touch with former ES11 students.

Based on the relatively small size of the Whitehorse community (approximate population 25,000), personal, community, and educational relationships that develop during the program's field studies and place-based activities give rise to frequent encounters with former students or their family in various environments. These conditions result in opportunities to engage with students and collect valuable data.

Postsecondary education

A summary of student education statistics indicates that 41 of the 45 interview respondents took part in postsecondary education and training but, of 150 students we followed through the postsecondary grants, 87 percent graduated from a university. Student participants had completed more than one program of studies and completed more than one degree. About one-fourth (25 percent) attended their first course at the local postsecondary institution — Yukon College — and then went to southern Canadian postsecondary institutions. Around one-fifth (20 percent) have attended university and training institutions in other countries (United Kingdom, United States, New Zealand, and Australia). Just slightly less than one-half (45 percent) of respondents have completed graduate studies — masters, PhD, or professional degrees. It is important to note that the Yukon student postsecondary attainment rate is low (approximately 24 percent). This program directly demonstrates the resiliency skills provided to support postsecondary engagement (Government of Canada 2014).

Employment

A summary of employment histories indicates that 85 percent of the students worked part-time in the Yukon during their time at studies and then took jobs in many Canadian and international locations. All respondents had jobs when they responded, and 20 percent indicated they were looking for other positions. The most interesting result, as it relates to this paper, is 45 percent held employment in fields related to environment or environmental monitoring.

Travel and recreational interests

A summary of travel and recreational interests indicates that all respondents have traveled to two or more continents, and all indicate an interest in more travel. Students have collectively visited or worked on every continent. All student respondents have remained active in outdoor activities and attribute many of their secondary school activities as the origins of these pursuits. About 40 percent of the respondents indicated they coach in areas related to their outdoor interests. Some former students listed a wide variety of outdoor activities, while others described their interests as outdoor activities and healthy lifestyles.

Personal background

Overall, almost three-quarters (70 percent), of the respondents indicated they were married or in long-term relationships; 55 percent have children and only 25 percent indicated they were single. Only one respondent had been divorced, after six years of marriage. Over one-half of the students considered themselves to be settled in their lifestyle, but of these, half indicated they were settled in their personal lives but not yet settled in their employment and also where they live geographically. About one-fourth (25 percent) of respondents live in the Yukon, and 60 percent of the respondents living outside the Yukon indicated they would like to live in the Yukon or in the province of British Columbia (south of the Yukon).

CITIZENSHIP SURVEY RESULTS

The second ongoing research project, which includes the intensive survey, paints a more comprehensive understanding of the impact the program had on long-term student engagement, resilience, and citizenship. To date, 37 of 45 surveys have been returned. The detailed survey includes a series of open-ended questions that address seven broad topics: education and training following high school, employment history, service to community or society, friendships persisting from high school, personal background, travel and recreational interests, and impacts of high school on subsequent choices and resilience. There also is a concluding open-ended section that invites further comments. It was in this final open-ended section that the theme of resiliency began to emerge through a spiral of cycles (Lewin 1947). The first four sections of questions provide quantitative information, while the last three sections of questions provide qualitative information. Overall, the intensive survey provides a blend of quantitative and qualitative information that yields insights into individuals' citizenship and the extent to which the ES11 program may have influenced subsequent life choices and resilience. Each of these topics is summarized in the next section.

Interim qualitative findings

To best represent the qualitative findings, we have chosen in this section of the analysis to provide representative quotes from the student responses that demonstrate key themes. Their voice, expressing common themes over a period of years, time and again provides compelling evidence regarding the values and subsequent actions of the former students.

Place-based learning impacts on student participants

Three quarters (75 percent) of the respondents indicated that the ES11 program impacted their subsequent career and educational choices. These quotes highlight the nature and magnitude of the impacts that the students, many of whom are Indigenous, drew from the key, life-changing experiences from participation in the place-based learning science program.

Male participant, 1998:

> The extended trip in ES was an invaluable learning experience for me regarding my ability to interact and communicate with others. As we (students) were tangled in socially intense learning environments (due to traveling with many adolescents), we were forced to learn and adapt to other people's behavior. My ability to communicate has continued to grow and is an integral part of my job. I started learning about tolerance, empathy, charity, and patience when I was in ES11, and it has allowed me to grow into the medical provider that I am today.

Female participant, 1996:

> Motivated, engaged, and challenging teachers with an awareness of current events reinforced my interest in global politics [and] encouraged my interest in development studies and sustainable communities. ... It had a lasting impact on my decision to pursue an education that included an environmental component.

Male participant, 1997:

> From the ES experience, I value opportunities to learn through practical experience and trying—and possibly failing. I value innovation. I value

environmental sustainability and policy decisions that have been informed by science.

Female participant, 2003:

Mapping (through GIS) and collecting salt from the side of the Alaska Hwy and taking it into the chemistry lab at Yukon College and actually figuring out how much salt was present in the gravel to link to caribou occurrences there, integrating chemistry, biology, and ecology. This experience as holistic learning changed how I viewed education and the world around me. I also learned that education could be fun and if I found something I enjoyed learning about, like how humans encounter their environment, it was up to me to figure it out for myself.

Male participant, 1996:

My high school experience cemented my interest in natural sciences. I sought out programs that could offer field-based studies.

Female participant, 1999:

My educational experience shaped my long-term employment goal to be part of an organization that promotes global sustainability, environmental awareness, and social conscience while building networks between governments and civil society. Ultimately, it reinforced my interest in development studies with a focus on environmental issues.

Cohort development and social networks

It is important to reiterate that students who took part in ES11 came from different schools and different communities. Even though they were only together as a cohort for a 93-day semester and many now live in different parts of the world, 60 percent of respondents indicated they remain in contact with friends they formed during the time they spent in ES11.

Male participant, 1997:

> I have maintained friendships with some people that I only met through this program and would have otherwise not known. Although not all people I kept in contact with, there are a select few that our relationship has grown over the years and I talk to on a regular basis (despite our geographical differences).

Male participant, 2001:

> I am very close to a few of the people from my ES class and continue significant relationships with them to this day. It is always a pleasure to see people from my ES class.

Service to community or society

The great majority of the students who participated in the program, 95 percent of the respondents, indicated they participate in service to their community or humanity. The following two quotes give a sense of this commitment to service and how service is related to citizenship and community resilience. Volunteer service is an essential aspect of being part of a community. It is an important consideration in how someone chooses to live because to be a volunteer encourages people to pursue activities they are passionate about. It also encourages people to be increasingly aware of issues outside their own social circles and provides the opportunity to foster attributes such as commitment and dedication. With such attributes, volunteers ensure sustainable programs that address local issues and needs.

Female student, 1999:

> I helped the Conservation Society of Sierra Leone facilitate community outreach workshops to encourage increased awareness of environmental and conservation of natural resource issues. I organize and conduct rapid assessment surveys of coastal sea-turtle habitat and local fishing practices.

Male student, 1998:

> One of the things I learned from ES11 is that everyone has a voice; it's all about how you say what you want to say. One of the most important things is to learn about the matter yourself and not rely on what others (e.g., pamphlets, protesters, etc.) are trying to convince you of. Do your own research, learn about it and you will have a stronger voice for it. People are more likely to listen to a knowledgeable person than a passionate, one-sided rant. And you might learn something yourself that changes your view of what others are saying. Being open to other people's opinions is as much a part of having a voice as knowing what to say.

Only two of the respondents indicated their voices were not heard and had little opportunity to influence decisions locally or nationally. A number of respondents spoke to problems with electoral systems, but they still felt the capacity to influence change. Every participant responded that they vote in civic, territorial, provincial, or national elections. This contrasts with the turnout of average Canadians as reported in the May 2011 national election; only 39 percent of Canadians aged 18–24 years and 45 percent aged 25–34 years voted (Mayrand 2012). These results once again reinforce the impact such programs have on student resilience and on community sustainability.

Open-ended section

The survey invited student respondents to make comments in an open-ended format. The following are a representative sample of their comments. Most of these relate to the experiential and PBE process.

Male student, 2004:

> After being in ES and after having gone through a moderate amount of post-secondary education, I think that experiential learning is a more robust way of learning and teaching. Being able to see a medial moraine, a U-shaped valley, the impacts of clear cutting, etc., turned "learning objectives" into concrete lessons. I think that in a perfect world all curriculums should/would be delivered in the same manner: intensive, tactile, and above all, meaningful.

Female student, 1995:

> Innovative experiential programs like these are an excellent model and should be expanded into other regions of Canada and other subject areas—perhaps in physics (engineering) and political/social studies? Decisions on where to move with my family and where to enroll my child for school will be heavily influenced by the availability of programs such as ES.

CONCLUSION

Recognizing the social and political context in which these students (Northern and Indigenous) often find themselves, the integrated experiential programs work to develop the human traits of criticality and resilience. These case studies demonstrate how educators strategically developed educational opportunities that supported Northern and Indigenous views of the interconnectedness of the environment that are reflected in resource stewardship. Through adversity and institutional resistance, this experiential program has chosen to support forms of resilience and emancipatory actions that include critical pedagogy (Kincheloe 2005). It is believed that through participating in the place-based learning active inquiry process, student motivation and engagement are heightened and ultimately lead to success. Place and educational tools such as GIS are foundational components in this process.

The "colorful five percent," as many northerners are often referred to, are similar to many Indigenous groups, which seem to have formed, by necessity or choice, a mantra of resilience to adversity that is often depicted in their history. Stories of northern survival, famine, government persecution, and the Gold Rush have spurred on many people to find significance in hardship.

The programs of study often use adversity as a motivating tool for personal and academic success. Students speak of the hardships they endure on extended field trips and the destruction of their environment. Like one young student said when he was up to his chest in cold water, "I love this stuff."

Teachers approve of the character-building qualities that such educational programs espouse:

> The experiential programs are set up in such a way that everybody gets a chance to fail. They find out their failing isn't a bad thing as long as he or

she uses it to learn something and then try it again. When I was talking [previously] about how we learn from our mistakes, I think the traditional school doesn't approach learning in this way. When a student makes a mistake in regular school, there is often a permanent consequence: now you got [a] 65 percent mark instead of [a] 75 percent mark, and that's there forever. That is not learning from your mistake, that is learning to not want to make a mistake, which to me is learning to not want to push your limit ... to provide an environment that is safe to go out and fail, because you can learn from it. ... And that's something everybody needs to learn, and I don't think a lot of people in our society learn that at all. ... Because I think our society stifles that. (Yukon teacher 2007)

Both case studies demonstrate similar dominant factors that promote success. These include supportive and committed staff, community involvement through the process of integration, inclusion of science and GIS technology as a means of engagement and critical inquiry, promoting issues of social justice, and the development of student qualities of resilience and strong sense of identity. Through outdoor field studies and environmental assessments supported by the use of GIS, students developed a heightened connection with the natural world that reflected cultural and community epistemologies.

The extent and nature of the responses to the detailed survey show a community of practice (Wenger 1998) of young adults from the periphery involved in community and a heightened understanding of place in active ways. They express the significant role this type of educational experience has had on subsequent life choices. Most participating students felt a sense of social resiliency and environmental responsibility, which we suggest are values and attitudes needed to address issues such as climate change.

Data from the quantitative and qualitative analysis shows complementary results. Those participating in the ES11 program demonstrated an uncommon level of resiliency, engagement, and civic and environmental responsibility. These students refer to the challenging and significant place-based field studies, the cooperative work relationships that develop during their semester, and diverse instructional processes used throughout the program as features that left lasting change. Field studies, including GIS, resonated with those students who learn best experientially and in social contexts. Students consistently reported the short- and long-term benefits attributed to their participation. A number of students indicated they struggled with conventional classes yet found success and engagement in the environmental field studies approach to courses. In terms of conventional academic scores, students in ES11 consistently outscored all other high school students taking similar courses.

Power of place-based education model

This place-based educational model helps create an educational climate in which students can develop into engaged, critical, and empowered learners of a diverse and complex globalized society (Bayne 2009). Pedagogical decisions are based on the teachers' insights into consciousness construction in the experience of themselves and their students, the interaction of the collective and the personal, diversity, social and educational theory, and instructional strategies. Such informed teaching creates unprecedented levels of awareness and higher forms of cognitive activity in the field of education.

One's experience of a place includes a complex combination of a specific physical environment, "our embodied encounter and the cultural ideas that influence the interpretations we make of the experience" (Wattchow and Brown 2011, ix). This provides rich potential for educators who are versed in place-based pedagogies. A student learning about the significance of a place, and how their beliefs and actions impact it, will be well positioned to reflect on how their civic actions may need to adapt to the challenges of the social, political, cultural, and economic environments. If the results of this research are indicative, far greater attention must be paid to the notion of place in education.

The development of citizens who internalize community and global challenges related to social and environmental goals appears to be an essential aspect of addressing phenomena related to climate change. This research sheds light on how a restructured public schooling may contribute to such development. In summary, the actions and values expressed by ES11 participants reflect those qualities of responsible citizenship needed to address the challenges identified by the IPCC Fifth Assessment Report (IPCC 2014). This research provides compelling quantitative and qualitative evidence indicating that educational processes involving place-based activities that encourage data collection (supported through GIS applications), reflection, and action are important antecedents to responsible and resilient citizenship.

REFERENCES

Bamberg, S., and G. Möser. 2006. "Twenty Years after Hines, Hungerford, and Tomera: A New Meta-Analysis of Psycho-Social Determinants of Pro-Environmental Behavior." *Journal of Environmental Psychology* 27: 14–25.

Barnhardt, C. 1999. "Standing Their Ground: The Integration of Community and School in Quinhagak, Alaska." *Canadian Journal of Native Education* 23 (1): 100–116.

Barr, S. 2003. "Strategies for Sustainability: Citizens and Responsible Environmental Behavior." *AREA* 35 (3): 227–240.

Bayne, G. U. 2009. "Joe L. Kincheloe: Embracing Criticality in Science Education." *Cultural Studies of Science Education* 4: 559–567.

Bloom, H. S., and R. Unterman. 2014. "Can Small High Schools of Choice Improve Educational Prospects for Disadvantaged Students?" *Journal of Policy Analysis and Management* 33 (2): 290–319.

Blue River Consulting. 2003. "Champagne & Aishihik First Nations' Upper Takhini River's Tributaries' JCS Investigations." Accessed January 20, 2017. http://yukonriverpanel.com/salmon/wp-content/uploads/2011/02/cre-54-02-takhini-river-tribs-jcs-investigations-report.pdf.

Chase, S., and B. McKenna. 2014. "Canada 'More Frank' about Climate Change." *The Globe and Mail*, June 9, 2014. www.theglobeandmail.com.

Creswell, J. W., and V. L. P. Clark, eds. 2011. *Designing and Conducting Mixed Methods Research*. Thousand Oaks, CA: SAGE.

Darling-Hammond, L., L. Flook, C. Cook-Harvey, B. Barron, and D. Osher. 2019. "Implications for Educational Practice of the Science of Learning and Development." *Applied Developmental Science*.

Darling-Hammond, L., P. Ross, and M. Milliken. 2006. "High School Size, Organization, and Content: What Matters for Student Success?" Brookings papers on education policy, 2006 (9): 163–203. Washington, DC: Brookings Institution Press.

David Suzuki Foundation. 2014. "IPCC Report Is Clear: We Must Clean Up Our Act." Accessed November 20, 2014. www.davidsuzuki.org.

Dewey, J. 1938. *Experience and education*. London: Collier-MacMillan.

Eccles, J. S., and R. W. Roeser. 2009. "Schools, Academic Motivation, and Stage-Environment Fit." *Handbook of adolescent psychology*. New York, NY: Wiley Publishing.

Emekauwa, E. 2004. "The Star with My Name: The Alaska Rural Systemic Initiative and the Impact of Place-Based Education on Native Student Achievement." Washington, DC: Rural School and Community Trust.

Felner, R. D., A. M. Seitsinger, S. Brand, A. Burns, and N. Bolton. 2007. "Creating Small Learning Communities: Lessons from the Project on High-Performing Learning Communities about 'What Works' in Creating Productive, Developmentally Enhancing, Learning Contexts." *Educational Psychologist* 42 (4): 209–221.

Fine, M., and J. I. Somerville, eds. 1998. "Small Schools, Big Imaginations: A Creative Look at Urban Public Schools." Chicago, IL: Cross City Campaign for Urban School Reform.

Freire, P. 1970. *Pedagogy of the Oppressed*. New York, NY: Continuum.

Gladden, J. 2001. "Wilderness Politics in Finnish Lapland: Core and Periphery Conflicts." *The Northern Review* 23: 59–81.

Glaser, E. M. 1985. "Critical Thinking: Educating for Responsible Citizenship in a Democracy." *National Forum: Phi Kappa Phi Journal* 65 (1): 24–27.

Government of Canada. 2014. "Education and Skills in the Territories." Accessed January 20, 2017. www.conferenceboard.ca/hcp/provincial/education/edu-territories.aspx.

Gruenewald, D. 2003. "The Best of Both Worlds: A Critical Pedagogy of Place." *Educational Researcher* 32 (4): 3–12.

Hines, J. M., H. Hungerford, and A. Tomera. 1986. "Analysis and Synthesis of Research on Responsible Environmental Behavior: A Meta-Analysis." *Journal of Environmental Education* 18 (2): 1–8.

Intergovernmental Panel on Climate Change (IPCC). 2014. "Climate Change 2014," AR5 Synthesis Report. www.ipcc.ch/report/ar5/syr.

Jamieson, G. S., C. D. Levings, B. C. Mason, and B. D. Smiley. 1999. "The Shorekeepers' Guide for Monitoring Intertidal Habitats of Canada's Pacific Waters." *Fisheries and Oceans Canada, Pacific Region*, modules 1, 2, and 3, vol. 1.

Kincheloe, J. 2005. *Critical Pedagogy*. New York, NY: Peter Lang.

Kleinfeld, J. S., G. W. McDiarmid, and D. Hagstrom. 1986. "Alaska's Small High Schools: Are They Working?" Fairbanks, AK: Center for Cross-Cultural Studies and Institute of Social and Economic Research, University of Alaska.

Lewin, K. 1947. "Group Decision and Social Change." In *Readings in Social Psychology*, edited by T. Newcomb and E. Hartley. New York, NY: Henry Holt.

Liepert, R. 2014. "Keystone XL Pipeline Will Keep Canada from Hitting 2020 Greenhouse Gas Emissions Targets, Say Critics." November 17, 2014, in *The Current*, Toronto, ON: CBC, produced by M. Tremonti. Radio broadcast episode. Accessed January 18, 2015. www.cbc.ca/thecurrent.

Louv, R. 2005. *Last Child in the Woods: Saving Our Children from Nature-Deficit Disorder*. Chapel Hill, NC: Algonquin Books of Chapel Hill.

Mayrand, M. 2012. "Declining Voter Turnout: Can We Reverse the Trend?" February 6, 2012. Elections Canada Online. Accessed January 5, 2015. www.elections.ca.

McInerney, Peter. 2011. "Coming to a Place Near You? The Politics and Possibilities of a Critical Pedagogy of Place-Based Education." *Asia-Pacific Journal of Teacher Education* 39 (1): 3–16.

O'Connor, K. 2009. "Puzzles Rather than Answers: Co-constructing a Pedagogy of Experiential, Place-Based and Critical Learning in Indigenous Education." Unpublished doctoral thesis, McGill University.

O'Connor, K. 2010. "Learning from Place: Re-shaping Knowledge Flow in Indigenous Education." *TRANS-Internet-Zeitschrift für Kulturwissenschaften*, no. 7/8–2. www.inst.at/trans/17Nr/8-2/8-2_oconnor.htm.

O'Connor, K., and R. Sharp. 2013. "Planting the Science Seed: Engaging Students in Place-Based Civic Actions." *European Scientific Journal* 4: 160–67.

Orr, D. W. 1994. *Earth in Mind: On Education, Environment, and the Human Prospect.* Washington, DC: Island Press.

Osher, D., and K. Kendziora. 2010. "Building Conditions for Learning and Healthy Adolescent Development: Strategic Approaches." In *Handbook of Youth Prevention Science*, edited by B. Doll, W. Pfohl, and J. Yoon. New York, NY: Routledge.

Penetito, W. 2009. "Place-Based Education: Catering for Curriculum, Culture and Community." *New Zealand Annual Review of Education* 18: 5–29.

Raffan, J. 1993. "The Experience of Place: Exploring Land as Teacher." *Journal of Experiential Education* 16 (1): 39–45.

Reeler, D. 2007. "A Three-fold Theory of Social Change and Implications for Practice, Planning, Monitoring and Evaluation." Community Development Resource Association.

Sharp, R. 1985. "Yukon Rural Education: An Assessment of Performance." Whitehorse, YT: Yukon Department of Education.

Sobel, D. 2004. "Place-Based Education: Connecting Classrooms and Communities." Great Barrington, MA: The Orion Society.

Stockholm Resilience Centre. 2017. "What Is Resilience?" Accessed January 20, 2017. www.stockholmresilience.org/research/research-news/2015-02-19-what-is-resilience.html.

Ten Dam, G., and M. Volman. 2004. "Critical thinking as a citizenship competence: Teaching strategies." *Learning and Instruction* 14: 359–79.

Tyack, D. B. 1974. *The One Best System: A History of American Urban Education* 95. Cambridge, MA: Harvard University Press.

Tyler, R. W. 1949. *Basic Principles of Curriculum and Instruction.* Chicago, IL: University of Chicago Press.

Wattchow, B., and M. Brown. 2011. *A Pedagogy of Place: Outdoor Education for a Changing World.* Clayton, Australia: Monash University Publishing.

Wenger, E. 1998. *Communities of Practice.* Cambridge, UK: Cambridge University Press.

Woodhouse, J. L., and C. E. Knapp. 2000. "Place-Based Curriculum and Instruction: Outdoor and Environmental Education Approaches." Charleston, WV: ERIC Clearinghouse on Rural Education and Small Schools.

→ CHAPTER 9

Engaging youth in spatial modes of thought toward social and environmental resilience

JASON DOUGLAS

INTRODUCTION

Structural inequities have historically situated low-income communities of color in uneven social and environmental circumstances (Grills et al. 2015; Hughes 2013). Current data illustrates the harmful effects of such inequities, revealing uneven distribution of green space (Loukaitou-Sideris 1993; Loukaitou-Sideris and Sideris 2009; Wolch et al. 2005), disproportionate educational outcomes (Potts 2003; Rumberger et al. 2012; Shriberg and Desai 2013), high rates of crime and violence (Dahlberg and Potter 2011; Howard 1996; Kellerman et al. 1998), and poor health outcomes (Brulle and Pellow 2006; Morello-Frosch et al. 2011) plaguing communities of color. South Los Angeles, California, presents a striking example of the detrimental effects of such inequities in and among communities of color. For example, uneven geographical distribution of green space has been linked health disparities, such as uneven rates of childhood obesity—approaching 39 percent in affected communities (Ogden et al. 2012). School suspension and expulsion rates upwards of 50 percent are also a serious concern (Gregory, Skiba, and Noguera 2010). Furthermore, in South LA, homicide is the leading cause of premature death for men aged 15–34. With a homicide rate of 38.8 for every 100,000 people of the population, triple the national homicide rate, the stress experienced by youth in underserved communities is exponentially higher than that in more affluent communities (LAPD 2014). The cumulative impacts of these injustices often produce a series of negative behavioral and psychological effects. That is, community members are exposed to social and environmental circumstances that influence individual- and community-level outcomes. For example, behavioral and psychological outcomes may include teenage pregnancy (Franklin et al. 1997; Wandersman and Florin 2003), political disenfranchisement (Flanagan and Levine 2010; Hope and Jagers 2014), and substance abuse (Sanders-Phillips et al. 2014; Welch-Brewer, Stoddard-Dare, and Mallett 2011).

In an effort to address negative behavioral and psychological effects, the concept of social and environmental resilience has come to the fore of research and practice, particularly concerning low-income communities of color. However, much of this research focuses on isolated areas of resilience (Khanlou and Wray 2014; Kidd and Davidson 2007). In response, more contextual research approaches are beginning to consider the connection between psychological and systems resilience, particularly with respect to promoting protective factors and reducing social and environmental vulnerability for youth attempting to cope with high stress environments that negatively affect youth physical and psychological health (Cohen and Schuchter 2013; Rew and Horner 2003).

SYSTEMS RESILIENCE

Systems resilience entails the healthy functioning of a system (hereafter referred to as *community*) that ultimately promotes an environment of positive youth development, economic growth, and environmental variability (for example, parks, libraries, and community centers). However, it should be noted that the resilience of a community does not necessarily reflect stability of ecological states, which could be potentially be detrimental to youth development. As such, community resilience is characterized by community ability to recover from disturbance (Adger 2000). For example, housing stability in the midst of public disinvestment provides an indicator of community resilience. Social resilience, on the other hand, is a somewhat nebulous concept with wide dissent across academic disciplines. Broadly speaking, social resilience is defined as an individual's ability to adapt to and cope with situations that present high levels of risk, stress, and adversity (Barankin and Khanlou 2007; Khanlou and Wray 2014; Luthar et al. 2000). Yet, social and environmental circumstances greatly differ by sociocultural and environmental characteristics, such as high homicide and broader crime rates. Therefore, when studying social and environmental vulnerabilities, one must begin to understand that what may be considered a vulnerability associated with poor educational outcomes for some may, in fact, be a protective factor within the climate of one's immediate community (Howard 1996). For example, gang membership is often associated with negative behaviors; however, gang membership may provide a sense of safety in communities plagued with high levels of violence.

A STRENGTHS-BASED APPROACH

Structural inequities that produce these negative outcomes in low-income communities and communities of color must be challenged through policy changes that address uneven access to quality education, job resources, community recreational resources, public transportation, and so on. It is simultaneously and equally important to develop a systematic understanding of how educators, academics, and the communities more broadly can promote protective factors that harness the resilience youth in marginalized communities have developed to cope with the harsh realities thrust upon them. Such research must go beyond said proxies of community vulnerabilities—for example, crime rates and demographic factors—toward grassroots knowledge production and action for reducing health-related social and environmental inequities and associated health disparities in low-income communities of color. These must include strength-based, social learning approaches that promote mental and physical health, foster social well-being, and reduce health disparities within the framework of preventive factors (Khanlou and Wray 2014; Smith et al. 2015).

STRUCTURAL INEQUITIES AND COMMUNITY FEATURES

It is critical to consider the multiple scales at which community youth experience myriad structural inequities. For example, the immediate experiences of youth and people of all ages directly influence development, wherein the way people think, interpret the world around them, and develop social and adaptive skills are interpreted at the individual (micro) level. It is particularly important to promote spaces and places fostering positive modes for understanding and working through the social and environmental contexts in which today's youth are situated.

At the family (macro level), youth are influenced by the strengths and challenges that their immediate families are faced with over time. The strengths essential to positive youth development are grounded in positive parent relations, open communication among family members, and support of extended family. Thus, families experiencing economic hardship, inequitable access to resources (for example, health care and healthy food), housing instability, and the like may be hindered from engaging in positive familial aspects that promote youth resiliency. Here it is noteworthy that these obstructions are often rooted in access to opportunities linked to immediate social and physical environments (described in the following section).

The immediate (neighborhoods) and extended (cities, localities, and so on) contexts represent the macro scale in which youth engage with the world around them. The social and environmental factors that surround youth and their families interact with individual- and family-level factors, influencing individual and family resilience. Some of these factors include access to educational and professional opportunities, access to resources that promote healthy living and afford a high quality of life, and policies and laws that mold social and built environments that afford high quality of life.

As mentioned above, many low-income communities of color live uneven realities, as reflected in the built environment. As such, thinking and acting spatially may offer an outlet for youth to begin to systematically and scientifically understand the unevenness of their everyday realities, offering a mode for youth to engage peers, family, community, and policy makers toward demanding policies that afford the same opportunities of their more affluent counterparts. These modes of engagement may build youth resilience in the forms of self- and collective efficacy (the ability to meet targeted goals individually and collectively) and psychological sense of community (feeling a sense of belonging to one's community). Ultimately, this may lead to establishing a broader sense of empowerment, which entails individual and community efforts to gain control over macro-level decisions that affect their everyday lives (Rappaport 1987; Speer et al. 2014; Zimmerman 2000).

PLACE AND PARTICIPATORY ACTION RESEARCH

Emerging evidence points to engagement in youth participatory action research (YPAR) as a promising collaborative research and social learning approach for promoting important aspects of youth resilience, including collective efficacy and empowerment (Berg et al. 2009; Cadell, Karanbanow, and Sanchez 2001; Douglas 2016; Ungar 2011). Youth participation in environmental justice research and action, in particular, has emerged as a powerful mode of youth engagement, providing a place-based approach to identifying youth social and environmental interests. These interests are frequently dubbed by policy makers and lobbyists as "community problems" that question the causal attributes of these interests, failing to explore social and environmental context.

For the purposes of this chapter, YPAR will be described as an iterative process entailing four phases—reflection, planning, research, and action. Through the YPAR phases described below, youth may go beyond asking why their environments are plagued with crime (reflection), for example, to working collaboratively with academics, community organizers, and the like to translate these observations and concerns into researchable questions and hypotheses. The hallmark of this approach is to design culturally situated research and education approaches for addressing these questions (planning), collecting data that reveals underlying causes related to community interests (research), identifying appropriate pathways for making collected data actionable (action), and discussing the challenges and accomplishments experienced in this process (reflection). This chapter accounts for micro and macro levels of resilience via process and outcome evaluation of a collaborative YPAR project involving ethnoculturally diverse high school and college students and educators living in Los Angeles, California.

Research context

The YPAR project took place over the course of 10 weeks, with 14 ethnoculturally diverse youth participating in an active research program concerned with social and environmental issues. All participants attended a continuation school in Los Angeles, California. The majority of the participants lived in South Los Angeles, an under-resourced area plagued by high crime and violence. Given the high school dropout rate upward of 50 percent in South LA communities (LADPH 2014), continuation schools present a viable alternative to traditional high schools for students who are considered to be "at risk" for not meeting graduation requirements. This at-risk designation may be attributed to a variety of reasons; however, the majority of students participating in this program attended the continuation school due to academic challenges, drug and alcohol use, and other negative behaviors exhibited in their former schools. Student participants ranged in age from 16 to 19 years of age. In an effort to improve upon the potential outcomes and collaborative nature of this project, eight undergraduate students from a nearby university participated as research assistants, collecting social and environmental data, as well as serving as mentors to the high school participants. This engaged and participatory approach was rooted in the tenets of resilience and empowerment, fostering interaction and collaboration among a range of stakeholders (Douglas 2016; Smith et al. 2015).

Critical reflection

To begin our interaction with the group in a YPAR context, we immediately entered into a phase of critical reflection to provide a space and place for the students to speak about their interests and learn about the community issues that concerned them most. On meeting both groups of students for the first time, I asked them if they would be willing to participate in an activity called the Geographical Imaginary, which presents an opportunity for students to begin to think spatially, considering the assets and deficits in their immediate communities, while being challenged to think about what their ideal neighborhoods would entail (Douglas 2016). Having previously employed this method in YPAR and community-based participatory research (CBPR) contexts, the Geographical Imaginary was tailored to foster youth engagement in the current project, affording a space for participants to express their immediate social and environmental interests, while thinking through the possibilities of what their neighborhoods could be through a spatial (neighborhood) lens. Some of the geographical realities expressed by the students included the following:

"Black-and-whites [police] left and right …"

"My neighborhood consists of loud people, busy streets, and criminal activity."

"There are some streets that are violent. Bad people in certain areas, drugs around the city, bunch of tagging crews, and some racism."

"The neighborhood I live in is always loud and dangerous."

"I live in a dirty and racist neighborhood."

The youth participating in this project expressed an overwhelming concern over the uneven neighborhood conditions in which they live. Such traits as criminal activity and corresponding police presence, drugs, violence, gang activity evidenced by "tagging" (graffiti), and noise were listed. These are neighborhood attributes frequently associated with social disorganization, wherein crime and subversive activity tend to be more prevalent in neighborhoods lacking social networks, collective efficacy, a sense of community, and a broader sense of neighborhood empowerment (Bursick 1998; Kubrin et al. 2011; Kubrin and Weitzer 2003; Shaw and McKay 1942). The students' experience of starting their school day being greeted by armed police, followed by going through a metal detector upon entering the school, further illustrates this social disorganization. It would logically follow that working with youth to develop individual and community resiliency pathways, particularly at micro and macro levels, may benefit from grassroots, participant-identified strategies and tactics to address social disorganization.

In response to these realities, the participants produced geographical imaginaries illustrating ideal neighborhoods comprising safe places to play and socialize. They expressed a desire for open spaces, parks, and a sense of community. This third concept is particularly striking, as it further indicated a need to address the social disorganization they experienced in their neighborhoods. Some of these quotes collected from the students illustrate ideal qualities they seek as a part of a community:

"Ice cream trucks, trees, parks, community resource centers, dog parks."

"A neighborhood where everyone cuts their lawn, green grass, plants, trees on the sidewalks."

"The perfect neighborhood is a neighborhood that's quiet and clean and has good neighbors and they all get along."

"My perfect neighborhood is kids running around, no criminal activity, parks, birds chirping, ice cream truck passes by."

"A place with new homes, quiet streets, and people in the community get along very well. A park at the corner that has a pond with ducks in the middle of the park. No tagging at all. Safety. Peace. A lot of grass and trees and soda machines. Around the area, it will have a church and bunch of restaurants."

These narrative excerpts highlight the importance of safe spaces conducive to youth socialization and development, as well as networks of people who "get along," which, according to social disorganization theory, may serve to improve neighborhood conditions (Bursick 1998; Kubrin et al. 2011; Kubrin and Weitzer 2003; Shaw and McKay 1942). However, it should be noted that improvements to the built environment—including recreational resources such as parks and community centers, and access to neighborhood amenities more broadly—should be accomplished by working with community to develop a sense of efficacy and community empowerment while promoting political engagement. Such changes would ideally occur at the policy level. Yet, it is essential to empower communities to demand policy change, a fundamental tenet of community resiliency (Berg et al. 2009; Maton 2008; Zimmerman et al. 2011).

Thinking through potential empowerment pathways, we entered into discussions of what we could do as a group to address some of their immediate concerns. Their impromptu responses articulated their distress over what they could achieve given their young age and the perceived importance of their interests. Furthermore, the high school participants were particularly troubled by the thought that local decision-makers didn't care about them. Yet, the fact that the students thought about the role of decision and policy makers to begin with is a testament to their awareness of social and environmental conditions, as well as their knowledge of power and influence in and around their communities. These fears were warranted, as the issues youth experience in their everyday lives frequently manifest on large and daunting scales, leading to a sense of disempowerment, wherein youth often express that workable solutions to their lived realities are out of reach (Chawla 1999; Hart 2008).

These real and imagined geographies set the stage for developing a research collaborative toward increasing youth understanding of the social and environmental circumstances in their locales through active and original YPAR. Participatory action research presents an opportunity for a range of stakeholders to engage in research toward achieving a common end. For our purposes (particularly given the short time frame we had in which to work with the students), the immediate goal was to expose disenfranchised youth to a rich learning opportunity, providing inroads to developing a sense of self- and collective efficacy, and a broader sense of civic engagement—all, hopefully, bridging a sense of empowerment for the participants to begin to place demands on their local schools, decision makers, and broader stakeholders to petition for equal access to resources. This initial exchange was rather telling, wherein the students expressed that our interaction had already transcended the dominant mode of engagement they had experienced with educators. By asking the students what they thought about their neighborhoods, inviting them to freely express their thoughts and feelings about the built environment on their own terms, the students conveyed a sense of belonging—the beginnings of a sense of trust and community through open and active communication. The students said, "So this is about us?" "You want to know what we think?" While troubling to hear the students describe a lack of familiarity with this type of engagement, it was encouraging to know that it appealed to them.

In this particular context, given the students' preference for neighborhoods that provide access to parks and broader nature experiences, while expressing a concern for high levels of violence in their communities, we agreed to launch a study of green space equity in their immediate locale. This experience would simultaneously provide a mode for the students to extend their spatial thinking, while grounding a situated understanding of recreational access in their neighborhoods in the concepts of space and place. Furthermore, this approach reflects inroads toward positive youth development (Grills 2002; Grills 2004; Gykye 1996) and social resiliency (Barankin and Khanlou 2007; Khanlou and Wray 2014).

Rolling up our sleeves

To begin to address some of the students' concerns, we moved into the second stage of YPAR—planning. In this, we collectively unpacked peer-reviewed literature regarding access to recreational space in Los Angeles in order to extend our contextual knowledge. One of these readings included Wolch et al. (2005), which describes the inequitable distribution of green space in Los Angeles by neighborhood racial characteristics. Our team reviewed pertinent parts of the publication in real time, paying particular attention to an analysis of the distribution of park acres in Los Angeles. We learned that per 1,000 people across the following ethnocultural neighborhoods (that is, areas in which one racial group comprises 75 percent or more of the population), there are 0.3 acres for Asian Pacific Islanders, 0.6 acres for Latinas/Latinos, 1.6 acres for African-Americans, and 31.8 acres for whites. These startling statistics were met with comments such as "That's not fair." By reviewing research regarding the local context, we were able to begin to garner a sense of the scope of inequities that youth in South LA and surrounding areas are faced with. Furthermore, this initial part of the planning process provided a mode for building a knowledge base through which to further reflect and ask questions of the local community.

In an effort to provide hands-on, engaging experiences for the students, we developed a workshop series for them to learn about conducting structured observations. In this activity, students left the classroom to observe their immediate school environment, drawing inferences on what they observed and how what they observed affected them directly. For example, the high school students observed a freeway overpass only three blocks away from their school. When asked how this may affect them, they immediately began to speak about how air pollution from car traffic impacted them. Furthermore, they began to draw connections to inequities associated with geographical location and exposure to toxic environments, speaking about how schools in other communities "probably" were not as close to these types of toxic environments. They began to ask questions about why they were exposed to such realities when those in more affluent communities less frequently experienced similar living circumstances. This led to students trying to frame this issue in a way that could convey their immediate interests.

And we did so, following up with lessons on how to develop concise and succinct research questions. Yet, asking students who have never been exposed to social and environmental research would gain little traction without training. Therefore, we developed activities for students to learn more about the process, while developing their own active and original research questions in context. The students began by identifying variables of interest, that is, exactly what they intended to measure. Some of the initial variables identified included communities, parks, violence, and gang activity, leading to their initial research question, "Do gang activity and drug use stop park visitation?" Working together, we eventually agreed to ask, "What is the relationship between neighborhood safety and access to recreational space in Los Angeles?" This question served as a starting point to begin to think about how to investigate this concerning phenomenon through rigorous research methods.

Since our research team wanted to explore a range of research methods to investigate the concerns identified throughout the reflection process, possibly involving human subjects, we also challenged the class to obtain institutional review board (IRB) certification—which is a fairly detailed research approval and documentation process, especially for research conducted with academic institutions of higher learning, like universities. This concept was completely alien to the students and school administrators, as none had been involved in research that could potentially lead to publications and policy suggestions. As our primary objective was to provide a platform for the students to understand how they can ethically participate in creating change through research, we decided it would behoove the project to involve the students in all steps salient to the research process. Promisingly, with the assistance of our undergraduate research assistants, who also were required to complete IRB training, all but one of the high school participants earned their certification. It was rather unexpected to observe a strong sense of accomplishment among the high school participants. Yet, they expressed that receiving official certification from an IRB-granting authority may provide opportunities that were previously thought to be out of reach, such as summer research internships.

Following IRB certification, we revisited our research question and learned about a range of methods that would enable us to begin to collect the data necessary to answer our question. Based on a series of lectures concerning quantitative and qualitative methods, the students decided it would be beneficial to administer community surveys to garner information about community access to recreational space. As the students lived in the service planning areas (SPA) of South LA (SPA 6) and South Bay (SPA 8), they thought it would be most beneficial to survey community members about the quality and accessibility of their local parks. Thus, the students reviewed survey instruments used in prior studies, deciding to design their own survey with the assistance of our undergraduate students to better reflect their research question. In addition, we used a structured observation tool designed by DeBate et al. (2011) to compare observed amenities and incivilities between 14 parks in South LA and 14 parks in the more affluent West LA (SPA 5) parks. As transportation proved challenging for high school student participation outside of their immediate neighborhoods, we agreed that the undergraduates would collect survey and observation data in the more affluent SPA 5, which served as a comparison area, also assisting with data collection in SPAs 6 and 8 (figure 9.1). In preparation for this phase, the high school and undergraduate students prepared Google maps of the parks where they intended to conduct research, detailing pertinent demographic characteristics associated with each area.

Before conducting structured observations and surveys, we invited the students to revisit the reflection process, offering their own thoughts and experiences on community parks. Some of the students expressed that they cannot go to their community parks, as they are associated with parts of their communities demarcated by gang territories. Crossing these lines would endanger their lives. Another student who frequently plays soccer in his local park had been robbed, while others had been chased, assaulted, and even shot. Not all but many of the students, particularly boys, described their local parks as spaces of violence, which invariably extends beyond

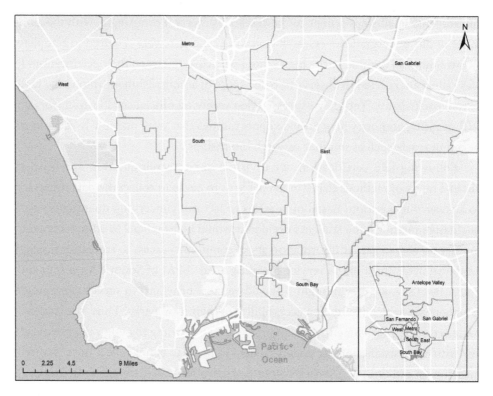

Figure 9.1. Los Angeles service planning areas.

Map courtesy of Jason Douglas. Data source: Los Angeles County GIS Data Portal.

park boundaries. To illustrate this, the students shared a story of how one of their classmates was beaten, almost to death, just outside the school grounds.

In an effort to give this some spatial grounding, we presented a participatory map developed with help from community members living in the Westmont/West Athens vicinity of South LA, which crossed over into the area where students were planning to conduct community surveys. The map was developed to illustrate community assets and deficits concerning recreational access in the Westmont/West Athens community. As the students reviewed the map and we discussed the purpose of participatory mapping, they immediately began to identify where they lived. This conversation instantaneously shifted to a discussion of gang territories, which was also a prominent feature of discussion among community youth who helped to identify map elements (for example, community assets and deficits). The students in the current study described gang territories, correcting some of the territories that I had initially marked with community input. They spoke of areas they could not visit because they wore the wrong colors (red or blue) or were not recognized in certain neighborhoods. They also pointed to the routes that could be safely taken to Helen Keller Park, shown in the southern portion of the map. Through this exercise, we developed a geographically situated understanding of student home ranges and local mobility patterns. This was extended by presenting tools used in related research projects, particularly spatial examples, which illustrated that the students' interests and concerns aligned with community members of various ages. Given this, they were eager to conduct community surveys and structured observations to see if the research tools they developed and adopted would work to further confirm that the community assets and deficits they had identified aligned with those of community members in their immediate locales.

Entering into the research phase, the high school and undergraduate students administered a total of 103 surveys in their community parks and surrounding areas between March and April 2015. For the high school students, this was, again, a new experience. Many of them were apprehensive about approaching people in their community, reflecting a concern that they would be perceived as "just kids." However, overcoming their anxiety, the high school participants collected an impressive number of surveys, excitedly returning to the classroom to share what they had learned during the course of their fieldwork. This was yet another important step toward harnessing youth resiliency through building self and collective efficacy. Furthermore, the students began to form a team focused on a common end, thus working toward a psychological sense of community.

As the students returned to the classroom to enter the data they had collected into a Microsoft Excel spreadsheet, revisiting the reflection phase of YPAR, they drifted into considerations of their everyday lived experiences, particularly concerning violence. During one of our data entry sessions, a student expressed concern that one of his friends had been shot the weekend before. Two more students spoke of a stabbing that happened on the street where they live, where the local media interviewed one of them. Taking this as another opportunity to ground our research, an undergraduate research assistant asked the high school participants to reflect on this further, particularly with respect to how community members respond to violence. Some of the high school students were concerned that "people only look out for themselves." Furthermore, they indicated that police presence increased tension in their neighborhoods. However, the high school students were adamant in advocating for "more of the right people on the street," which aligns with social disorganization theory, wherein strong community networks and community empowerment help to curb community crime. This further grounded our understanding that promoting social and environmental resilience would be hinged on the principles of empowerment.

As we moved into the data analysis portion of our project, the students were rewarded by seeing two weeks of data entry come together in the form of an Excel spreadsheet. We would have preferred to use advanced data analysis software, but working with students in a limited capacity warranted using tools that they were already familiar with, as our objective was to move toward developing a more situated understanding of community conditions concerning recreational equity. To make the activity engaging, we started out by challenging the class to hypothesize how the surveyed communities responded to individual survey questions. For example, we challenged the students to identify which communities had the highest perceived crime rates. Interestingly, they unanimously said that Black and Latino communities would be highest, and they were correct. However, they were quite perplexed to find that Asian and White community members logged more responses, indicating their local parks and recreational resources were littered. This, and other data points, led to the students questioning why these differences occur. For example, one student exclaimed, "That's not right," leading to the class hypothesizing that perhaps some communities perceive their spaces differently, wherein higher crime rates and unfortunate community conditions stemming from systemic inequities plaguing Black and Brown communities have become normative, manifesting as an accepted facet of lived reality. Furthermore, as the perception of neighborhood crime, statistically speaking, was not significantly different, the students explained that we need to consider specific types of crime, as violent crime in communities of color may be perceived as more significant than lesser crimes of litter and theft.

The class was also interested to learn that men in their communities were more concerned with safety issues in local parks than women. The students discussed this curious finding related to community violence, particularly gang violence—which they noted often affects men disproportionately. In fact, all the violent in-park incidents that the students mentioned in their own experiences concerned men and boys. Also, it was found that Black and Brown community members invariably thought that parks were dangerous at night and Latina/Latino community members ranked parks as places that could be dangerous for children. This was further reflected in the qualitative responses to a survey question probing for park deficits, wherein Black and Brown community members invariably indicated that gang activity and violence are major problems, a significant difference when compared with responses from White and Asian survey participants.

Our structured observations revealed that parks in SPA 6 and 8, while having similar amenities to parks in SPA 5 (figure 9.2), had more undesirable conditions, including litter, evidence of drug and alcohol use, and dirty bathrooms. Furthermore, their observations also revealed that play equipment in SPA 5 parks tended to be of a higher quality, with more elements designed to foster youth development, such as various textures and climbing equipment. Unfortunately, this did not come as a surprise to the high school students, who again reflected on the poor conditions of their local parks.

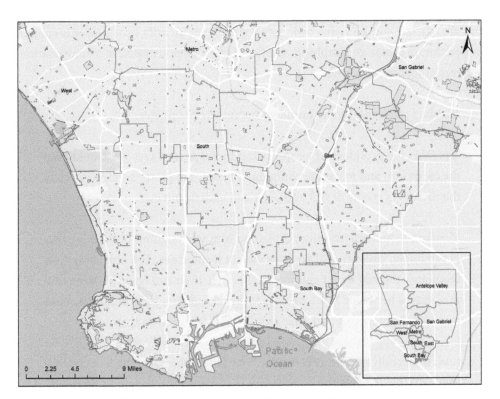

Figure 9.2. Los Angeles green space distribution by service planning area.

Map courtesy of Jason Douglas. Data source: Los Angeles County GIS Data Portal.

REFLECTING ON ACTIONABLE PRACTICES AND OUTCOMES—THE ROLE OF POWER

As the class ended, leaving little time to take action on what we had learned, we reflected on possibilities for using the data that had been collected with this participatory process. In this discussion, the high school and undergraduate students explored possibilities at various scales, from federal to city. For example, one high school student thought we should identify how to share project outcomes with President Obama. This was immediately met with skepticism from some of his classmates, who responded with statements including "He doesn't have time to worry about parks" and "How would we reach him?" While some of the students thought this approach might be farfetched, they were able to identify how power is critical in determining their geographical realities. This is demonstrated in a conversation between students, as follows:

Marvin: I think mostly the president because he has power.

Karl: He's the president. If he speaks out to people and says we should keep the park clean and make the community better, I'm pretty sure everyone in the country is gonna hear it and see it and understand it.

Marvin: I would tell him that, explain to him what's going on in our parks. For example, violence. … Like how violent it could get. Tell him about how unsanitary it could be for children in the park. If we can bring children into this, maybe he would listen because he has daughters. He wouldn't want to take his daughters to a park that was dirty, violent, and dangerous.

Karl: If he really loves this country… The bigger the park, the less violence there will be. It helps the community. Talking to the people in the neighborhood, it's a place to build up friendship and reduce crime.

Marvin: We should take this data to city hall or the people that work in the recreational centers because I am pretty sure they have a say of what gets built in the park and what gets left out.

This conversation illustrated an understanding of power; that is, how to gain real influence to address the adversities communities face through modifying the built environment. They included theory- and practice-based elements, a praxis-based approach they had learned about while engaging in the research process, from an evidence base indicating the role of green space and park programming in reducing crime (Medway et al. 2016; Weiss et al. 2011) to the role of these spaces in promoting a sense of community (Floyd et al. 2009; Wolch et al. 2014). Interestingly, they also identified methods for appealing to decision makers' personal feelings by speaking of the detrimental effects of uneven neighborhood conditions on community youth. In this, it appeared that the students were moving toward a mode of thought exemplified by community organizers, who work with communities to demand policy changes toward improving social and environmental conditions (Cheezum et al. 2013; Minkler and Wallerstein 2012; Speer et al. 2014).

Creating safe space

As an important part of this participatory research process, we realized that it would be important to have a space and place for starting this type of interaction. Because the research that we produced together was somewhat of an anomaly for the students and the school, the students thought it would be essential to both provide a space and place for the students to take action in their communities and introduce social justice and community organizing curricula in their current school programming. This recognition illustrated a desire to take action to better parks and broader neighborhood conditions and also develop inroads to this action through their school environment. The students' thinking on this matter is particularly encouraging, as they knew that change was necessary, and they needed a space through which to initiate change.

In a culminating discussion with the students, we explored what they learned and how they felt during the course of our project. The aforementioned desire to produce actionable data and ensuing social and environmental change was particularly encouraging. However, the students also expressed additional feelings of growth grounded in the tenets of empowerment and resilience. The following real-time discussion thread concerning what the students were most surprised about in our project illustrates some pertinent aspects of resilience:

Shawn: Confidence in the way of being more open and being able to go out and get surveys done. Like being confident to be able to speak to people.

Hector: It was okay. I thought it was kind of a shock with the surveys, and I never thought I would do something like that. The outcome was pretty good because, like what Shawn said, it helped build confidence in talking to people.

Kanai: … how they [undergraduate students] told us that we had a voice in changing the community, like with the parks and all that. I found that interesting, because I didn't know that.

These statements indicate a growing sense of self-efficacy, wherein the students understood that, regardless of their age, they had the ability to achieve something beyond the expectations of school educators, community members, and society more broadly. That is, they found that using scientific methods to develop a situated understanding of an issue, access to recreational space in this context, effectively amplified their own voice in changing the community. For their communities, their developing voice targeted changing the "bad things in the park—gang violence, drug violence, homelessness, police brutality." These experiences that had been accepted as part of an everyday lived reality were transformed into an understanding of systemic inequities that disproportionately affect low-income communities of color, and the research team was poised to take action with the support of their school community, expressing a sense of empowerment.

Yet, it is critical to note that this growing resilience and sense of empowerment were fundamentally linked to the social learning partnership formed during our brief collaboration. In this, the students began to express a sense of collective efficacy, wherein they realized that the strength they were gaining was bolstered by the power of an ethnoculturally and age-diverse group of people with common interests. This, in turn, further promoted social resilience through a research collaborative concerning inequitable access to green space. The students' sentiments on this matter were promisingly expressed in the following statements:

Mauricio: Since, like, they [undergraduate students] are already in college, you know, and they come to the school to help us out. So, like, they're teaching us what they know, so later on in life when we get there, we'll know the stuff that we already know.

Kanai: Normally I wouldn't have done the surveys, but since you were doing it and everyone else was doing it, it only seemed right to do it. We're not all doing it alone, so, you know.

They also went on to say:

Mauricio: [working with college students] … inspired me to do good in school. The way you guys are. Getting an education, making life kind of simple. It kind of showed me that I can do that, too, instead of being lazy and not doing anything. So, since I've been in this class, I've been doing good.

Hector: Before I was in this class, my grades were just downhill. But, just having to do what we did in here, I was just like, oh, I should do that in all my classes. It helped me to get my grades up and keep doing it.

Maria: I'm going to have to agree with him, because that's the same thing I was thinking.

These statements illustrate the power of social learning grounded in collaborative activities toward promoting social and environmental resilience. The research team understood that working with peers and mentors toward a common goal would help them better address their goal of systematizing their knowledge of recreational access, eventually growing into life skills fundamental to instilling a sense of resiliency. Furthermore, the immediate effects included self-reported improved educational outcomes and a growing demand for progressive research and action programs for the high school students to continue to explore their interests and take action.

CONCLUSION

Working with youth in a participatory capacity can promote social and environmental resilience and community empowerment pathways toward advocating for social and environmental equity. The youth participants described in this chapter exemplified many of the promising aspects of this approach. Working together to identify community interests (reflection), develop approaches to understand these interests (planning), investigate them (research), and mobilize (action) to address them engendered a developing sense of resilience and psychological empowerment throughout the group. In this approach, the youth participants rescaled large, daunting issues to the local context, recalibrating their awareness of the issues to their own interests and producing a realization of their role in actively understanding and redressing systemic inequities that result in inequitable neighborhood outcomes. In their explorations of these inequities, they also collaboratively identified empowerment pathways that would be immediately beneficial in their call for political voice and power, and formalized structures in the school environment for actively promoting these empowerment pathways. This last point, perhaps most promisingly, revealed the group's realization that the primary neighborhood institution immediately affecting their lives is their school. Thus, the call for recalibrating the school environment—a space intended for learning and positive youth development—to promote social and environmental action was revealing. It will behoove institutions and the neighborhoods they serve to actively engage community youth in identifying empowering educational practices.

Engaging in spatial modes of thought supported the development of a situated understanding of social and environmental inequities, which were then relocated to the local context concerning the intersection of crime and access to recreational space in South LA low-income communities of color. This spatial mode of thought accordingly provided the impetus for moving beyond the daunting sphere of public policy to engagement in active research toward promoting social and environmental equity. In this pursuit, a group of young scholars emerged, poised to take action with newfound knowledge and tools that, in the context of collective action, revealed a developing sense of self- and collective efficacy and a psychological empowerment. These emergent, community-engaged scholars thus moved toward social and environmental resilience in their collective pursuit of knowledge, action, and community power.

REFERENCES

Adger, N. W. 2000. "Social and Ecological Resilience: Are They Related?" *Progress in Human Geography* 24 (3): 347–64.

Barankin, T., and N. Khanlou. 2007. *Growing Up Resilient: Ways to Build Resilience in Children and Youth. Center for Addiction and Mental Health.* ISBN 978-0-88868-504-9.

Berg, B., E. Coman, and J. Schensul. 2009. "Youth Action Research for Prevention: A Multi-level Intervention Designed to Increase Efficacy and Empowerment Among Urban Youth." *American Journal of Community Psychology* 43 (3–4): 345-59. https://doi.org/10.1007/s10464-009-9231-2.

Berman, S. L., W. M. Kurtines, W. K. Silverman, and L. T. Serafini. 1996. "The Impact of Exposure to Crime and Violence on Urban Youth." *American Journal of Orthopsychiatry* 66 (3): 329–336. https://doi.org/10.1037/h0080183.

Brulle, R.T J., and D. N. Pellow. 2006. "Environmental Justice: Human Health and Environmental Inequalities." *Annual Review of Public Health* 27: 103–24.

Bursick, J. 2017. "The Role of Litigation in Pursuing Environmental Justice in Richmond, CA." *Environmental Justice Litigation*, Spring 2017: 1–19.

Cadell, S., J. Karanbanow, and M. Sanchez. 2001. "Community, Empowerment, and Resilience. Paths to Wellness." *Canadian Journal of Community Mental Health* 20: 21–35.

Chawla, L. 1999. "Life Paths Into Effective Environmental Action." *The Journal of Environmental Education,* 31 (1): 15–26. https://doi.org/10.1080/00958969909598628.

Cheezum, R. R., C. M. Coombe, B. A. Israel, R. J. McGranaghan, A. N. Burris, S. Grant-White, A. Weigl, and M. Anderson. 2013. "Building Community Capacity to Advocate for Policy Change: An Outcome Evaluation of the Neighborhoods Working in Partnership Project in Detroit." *Journal of Community Practice* 21 (3): 228–47. https://doi.org/10.1080/10705422.2013.811624.

Cohen, A. K., and J. W. Schuchter. 2013. "Revitalizing Communities Together." *Journal of Urban Health* 90 (2): 187–96. https://doi.org/10.1007/s11524-012-9733-3.

Comber, A., T. Boeck, D. Montfort, J. Hardman, C. Jarvis, and P. Kraftl. n.d. "Methods for the Spatial Analysis of Community Wellbeing, Resilience and Vulnerability." https://agile-online.org/conference_paper/cds/agile_2012/proceedings/papers/paper_comber_methods_for_the_spatial_analysis_of_community_wellbeing_resilience_and_vulnerability_2012.pdf.

Dahlberg, L. L., and L. B. Potter. 2001. "Youth Violence Developmental Pathways and Prevention Challenges." *American Journal of Preventive Medicine* 20 (1): 3–14. https://doi.org/10.1016/S0749-3797(00)00268-3.

Daining, C., and D. DePanfilis. 2007. "Resilience of Youth in Transition from Out-of-Home Care to Adulthood." *Children and Youth Services Review* 29 (9): 1158–78. https://doi.org/10.1016/j.childyouth.2007.04.006.

DeBate, R. D., E. J. Koby, T. E. Looney, J. K. Trainor, M. L. Zwald, C. A. Bryant, and R. J. McDermott. 2011. "Utility of the Physical Activity Resource Assessment for Child-centric Physical Activity Intervention Planning in Two Urban Neighborhoods." *Journal of Community Health* 36: 132–40. https://doi.org/10.1007/s10900-010-9290-1.

Douglas, J. A. 2016. "What's good in the 'hood: The production of youth, nature, and knowledge." In L. Shillington and A. M. Murnaghan (eds.), *Children, Nature, and Cities.* Burlinton, VT: Ashgate Publishing Company.

Flanagan, C., and P. Levine. 2010. "Civic Engagement and the Transition to Adulthood." *The Future of Children* 20 (1): 159–79.

Floyd, M. F., W. C. Taylor, and M. Whitt-Glover. 2009. "Measurement of Park and Recreation Environments that Support Physical Activity in Low-Income Communities of Color: Highlights of Challenges and Recommendations." *American Journal of Preventive Medicine* 36 (4): S156–S160. https://doi.org/10.1016/j.amepre.2009.01.009.

Forrest-Bank, S., N. Nicotera, E. Anthony, B. Gonzales, and J. Jenson. 2014. "Risk, Protection, and Resilience Among Youth Residing in Public Housing Neighborhoods." *Child & Adolescent Social Work Journal* 31 (4): 295–314. https://doi.org/10.1007/s10560-013-0325-1.

Franklin, C., D. Grant, J. Corcoran, P. O'Dell Miller, and L. Bultman. 1997. "Effectiveness of Prevention Programs for Adolescent Pregnancy: A Meta-Analysis." *Journal of Marriage and Family* 59 (3): 551–67. https://doi.org/10.2307/353945.

Gregory, A., R. J. Skiba, and P. A. Noguera. 2010. "The Achievement Gap and the Discipline Gap: Two Sides of the Same Coin?" *Educational Researcher* 39 (1): 59–68. https://doi.org/10.3102/0013189X09357621.

Grills, C. 2002. "African-centered psychology: Basic principles." In T. Parham (ed.), *Counseling Persons of African Descent: Raising the Bar of Practitioner Competence* 10–25. Thousand Oaks, CA: Sage.

Grills, C. 2004. "African psychology." In R. Jones (ed.), *African Psychology* 171–208. Hampton, VA: Cobb and Henry.

Grills, C., D. Cooke, J. A. Douglas, S. Villanueva, A. M. Subica, and B. Hudson. 2015. "Culture, Racial Socialization, and Positive African American Youth Development." *Journal of Black Psychology*, 0095798415578004. http://doi.org/10.1177/0095798415578004.

Gyekye, K. 1996. *African Cultural Values: An Introduction.* Accra, Ghana: Sankofa.

Hart, R. A. 2008. "Stepping Back from 'The Ladder': Reflections on a Model of Participatory Work with Children." In A. Reid, B. B. Jensen, J. Nikel, V. Simovska (eds.), *Participation and Learning.* Springer, Dordrecht. https://doi.org/10.1007/978-1-4020-6416-6_2.

Hope, E. C., and R. J. Jagers. 2014. "The Role of Sociopolitical Attitudes and Civic Education in the Civic Engagement of Black Youth." *Journal of Research on Adolescence* 24 (3).

Howard, D. E. 1996. "Searching for Resilience among African-American Youth Exposed to Community Violence: Theoretical Issues." *Journal of Adolescent Health* 18 (4): 254–62. https://doi.org/10.1016/1054-139X(95)00230-P.

Hughes, G. 2013. "Racial justice, hegemony, and bias incidents in U.S. higher education." *Multicultural Perspectives* 15: 126–32. https://doi.org/10.1080/15210960.2013.809301.

Kellerman, A. L., D. S. Fuqua-Whitley, F. P. Rivara, and J. Mercy. 1998. "Preventing Youth Violence: What Works?" *Annual Review of Public Health* 19: 271–92.

Khanlou, N. 2007. "Youth and post-migration cultural identities: Linking the local to the global." In A. Asgharzadeh, E. Lawson, K. U. Oka, and A. Wahab (eds.) *Diasporic Ruptures*. Sense Publishers.

Khanlou, N., and R. Wray. 2014. "A Whole Community Approach toward Child and Youth Resilience Promotion: A Review of Resilience Literature." *International Journal of Mental Health and Addiction* 12 (1): 64–79. https://doi.org/10.1007/s11469-013-9470-1.

Kidd, S. A., and L. Davidson. 2007. "You Have to Adapt Because You Have No Other Choice: The Stories of Strength and Resilience of 208 Homeless Youth in New York City and Toronto." *Journal of Community Psychology* 35 (2): 219–38. https://doi.org/10.1002/jcop.20144.

Kubrin, C. E., G. D. Squires, S. M. Graves, and G. C. Ousey. 2011. "Does Fringe Banking Exacerbate Neighborhood Crime Rates?" *Criminology and Public Policy* 10: 437–66.

Kubrin, C. E., and R. Weitzer. 2003. "New Directions in Social Disorganization Theory." *Journal of Research in Crime and Delinquency* 40 (4): 374–402.

Liebenberg, L., M. Ungar, and F. V. d. Vijver. 2011. "Validation of the Child and Youth Resilience Measure-28 (CYRM-28) Among Canadian Youth." *Research on Social Work Practice* 22 (2): 219–226. https://doi.org/10.1177/1049731511428619.

Los Angeles Police Department. 2014. LAPD Statistical Data. www.lapdonline.org/statistical_data.

Loukaitou-Sideris, A. 1993. "Privatisation of Public Open Space." *Town Planning Review* 64 (2): 139–68.

Loukaitou-Sideris, A., and A. Sideris. 2009. "What Brings Children to the Park? Analysis and Measurement of the Variables Affecting Children's Use of Parks." *Journal of the American Planning Association* 76 (1): 89–107.

Luthar, S. S. 1991. "Vulnerability and Resilience: A Study of High-Risk Adolescents." *Child Development* 62 (3): 600–616.

Luthar, S. S. 2003. *Resilience and Vulnerability: Adaptation in the Context of Childhood Adversities*. Cambridge, UK: Cambridge University Press.

Luthar, S. S., and A. Goldstein. 2004. "Children's Exposure to Community Violence: Implications for Understanding Risk and Resilience." *Journal of Clinical Child and Adolescent Psychology: The Official Journal for the Society of Clinical Child and Adolescent Psychology.* American Psychological Association, Division 53, 33(3): 499–505. https://doi.org/10.1207/s15374424jccp3303_7.

Luthar, S. S., C. H. Doernberger, and E. Zigler. 1993. "Resilience is Not a Unidimensional Construct: Insights from a Prospective Study of Inner-City Adolescents." *Development and Psychopathology* 5 (4): 703–17. https://doi.org/10.1017/S0954579400006246.

Luthar, S. S., and D. Cicchetti. 2000. "The Construct of Resilience: Implications for Interventions and Social Policies." *Development and Psychopathology* 12 (4): 857–85. https://doi.org/10.1017/S0954579400004156

Luthar, S. S., D. Cicchetti, and B. Becker. 2000. "The Construct of Resilience: A Critical Evaluation and Guidelines for Future Work." *Child Development* 71 (3): 543–62.

Luthar, S. S., and S. H. Barkin. 2012. "Are Affluent Youth Truly 'At Risk'? Vulnerability and Resilience across Three Diverse Samples." *Development and Psychopathology* 24 (2): 429–49. https://doi.org/10.1017/S0954579412000089.

Maton, K. I. 2008. "Empowering Community Settings: Agents of Individual Development, Community Betterment, and Positive Social Change." *American Journal of Community Psychology* 41 (1–2): 4–21.

Medway, D., C. Parker, and S. Roper. 2016. "Litter, gender and brand: The anticipation of incivilities and perceptions of crime prevalence." *Journal of Environmental Psychology* 45: 135–44.

Minkler, M., and N. Wallerstein. 2012. "Introduction to community organizing and community building." In M. Minkler, (ed.), *Community Organizing and Community Building for Health and Welfare, 3rd ed.* 5–26. New Brunswick, NJ: Rutgers University Press.

Morello-Frosch, R., P. Brown, M. Lyson, A. Cohen, and K. Krupa. 2011. "Community Voice, Vision, and Resilience in Post-Hurricane Katrina Recovery." *Environmental Justice* 4 (1): 71–80. https://doi.org/10.1089/env.2010.0029.

Ogden, C. L., M. D. Carroll, B. K. Kit, and K. M. Flegal. 2012. "Prevalence of obesity and trends in body mass index among U.S. children and adolescents, 1999-2010." *JAMA* 307 (5) (2012): 483–90. https://doi.org/10.1001/jama.2012.40.

Potts, R. G. 2003. "Emancipatory Education Versus School-Based Prevention in African American Communities." *American Journal of Community Psychology* 31: 173–83. https://doi.org/10.1023/A:1023039007302.

Rappaport, J. 1987. "Terms of empowerment/exemplars of prevention: Toward a theory for community psychology." *American Journal of Community Psychology* 15: 121–48. https://doi.org/10.1007/BF00919275.

Rew, L., and S. D. Horner. 2003. "Youth Resilience Framework for Reducing Health-Risk Behaviors in Adolescents." *Journal of Pediatric Nursing* 18 (6): 379–88. https://doi.org/10.1016/S0882-5963(03)00162-3.

Rumberger, R., and S. Rotermund. 2012. "The Relationship between Engagement and High School Dropout." *Handbook of Research on Student Engagement*, edited by S. L. Christenson, A. L. Reschly, and C. Wylie, 491–513.

Sanders-Phillips, K., W. Kliewer, T. Tirmazi, V. Nebbutt, T. Carter and H. Key. 2014. "Perceived Racial Discrimination, Drug Use, and Psychological Distress in African American Youth: A Pathway to Child Health Disparities." *Journal of Social Issues* 70 (2): 279–97. https://doi.org/10.1111/josi.12060.

Shaw, C. R., and H. D. McKay. 1942. *Juvenile Delinquency and Urban Areas*. Chicago, IL: University of Chicago Press.

Shriberg, D., and P. Desai. 2013. "Bridging Social Justice and Children's Rights to Enhance School Psychology Scholarship and Practice." *Psychology in the Schools* 51 (1): 3–14. https://doi.org/10.1002/pits.21737.

Smith, J. G., B. DuBois, and M. E. Krasny. 2015. "Framing for Resilience through Social Learning: Impacts of Environmental Stewardship on Youth in Post-Disturbance Communities." *Sustainability Science* 11 (3): 441–53. https://doi.org/10.1007/s11625-015-0348-y.

Speer, P. W., E. A. Tesdahl, and J. F. Ayers. 2014. "Community organizing practices in a globalizing era: Building power for health equity at the community level." *Journal of Health Psychology* 19 (1): 159–69.

Towe, V., A. Chandra, J. Acosta, R. Chari, L. Uscher-Pines, and C. Sellers. 2015. *Community Resilience Learn and Tell Toolkit*. Accessed May 2, 2016. www.rand.org/pubs/tools/TL163.html.

Ungar, M. 2004. "A Constructionist Discourse on Resilience: Multiple Contexts, Multiple Realities among At-Risk Children and Youth." *Youth & Society* 35 (3): 341–65. https://doi.org/10.1177/0044118X03257030.

Ungar, M. 2011. "The Social Ecology of Resilience: Addressing Contextual and Cultural Ambiguity of a Nascent Construct." *American Journal of Orthopsychiatry* 81 (1): 1–17. https://doi.org/10.1111/j.1939-0025.2010.01067.x.

Wandersman, A., and P. Florin. 2003. "Community Interventions and Effective Presentation." *American Psychologist* 58 (6–7): 441–48.

Weiss, C. C., M. Purciel, M. Bader, J. Quinn, G. Lovasi, K. M. Neckerman, A. G. Rundle. 2011. "Reconsidering access: Park facilities and neighborhood disamenities in New York City." *Journal of Urban Health* 88: 297–310. https://doi/org/10.1007/s11524-011-9551-z.

Welch-Brewer, C. L., P. Stoddard-Dare, and C. A. Mallett. 2011. "Race, Substance Abuse, and Mental Health Disorders as Predictors of Juvenile Court Outcomes: Do They Vary by Gender?" *Child and Adolescent Social Work Journal* 28 (3): 229–41.

Wells, N. M. 2014. "The Role of Nature in Children's Resilience: Cognitive and Social Processes." *Greening in the Red Zone*, edited by K. G. Tidball and M. E. Krasny 95–109. Springer Netherlands. https://doi.org/10.1007/978-90-481-9947-1_7.

Wolch, J., J. P. Wilson, and J. Fehrenbach. 2005. "Parks and Park Funding in Los Angeles: An Equity- Mapping Analysis." *Journal of Urban Geography* 26 (1): 4–35.

Wolch, J., J. Byrne, and J. P. Newell. 2014. "Urban Green Space, Public Health and Environmental Justice: The Challenge of Making Cities 'Just Green Enough.'" *Landscape and Urban Planning* 125 (May): 234–44.

Zimmerman, M. A. 2000. "Empowerment theory: Psychological, organizational, and community levels of analysis." In J. Rappaport and E. Seidman (eds.), *Handbook of Community Psychology*, 44–59. New York: Plenum Publishers.

Zimmerman, M. A., S. E. Stewart, S. Morrel-Samuels, S. Franzen, and T. M. Reischl. 2011. "Youth empowerment solutions for peaceful communities: Combining theory and practice in a community-level violence prevention curriculum." *Health Promotion & Practice* 12 (3): 425–39. https://doi.org/10.1177/1524839909357316.

→ CHAPTER 10

Health, place, and space: Public participation GIS for rural community power

**SHEILA LAKSHMI STEINBERG AND
STEVEN J. STEINBERG**

INTRODUCTION

Around the world, there is a strong connection between people's health and the places they live. Perhaps nowhere else is this connection more apparent than in rural, agricultural communities where the daily interaction between people and their environment is the basis of both individual and community success. These communities may frequently face challenges, both anticipated and unexpected, and must work together to be resilient. Why? Because they must learn to navigate environmental changes and challenges often with the limited social capital, economic tools, and local resources available within the community. In this chapter, we focus on rural communities in two California counties where community members' health was believed to be negatively impacted by the same agricultural industries that are so essential to the success of the communities and provide fresh fruits and vegetables to the global supply chain.

In this chapter, we examine the experience of rural farmer worker communities in Monterey County and Tulare County, California (figure 10.1). While many think of California as the home of urbanized areas such as San Francisco and Los

Angeles

SHEILA LAKSHMI STEINBERG AND STEVEN J. STEINBERG

Angeles, filled with traffic, a lesser-known fact is that 80 percent of the state is actually defined as rural (State of California 2020). People living and working in these rural places are self-reliant and accustomed to helping one another. But this does not mean that these rural communities exist as a picturesque pastoral scene from a Hollywood movie. Many people engaged in jobs in rural communities are tied to a natural resource-based economy such as agriculture, forestry, mining, or fishing, each coupled with a wide variety of risks and dangers. In this chapter, we focus on a spatial understanding of how six agricultural communities experienced polluted environments resulting, in part, from pesticide drift and exposure.

Figure 10.1. The locations of Monterey County, along the Central California coast, and Tulare County in California's Central Valley.

Map courtesy of Steven Steinberg and Sheila Steinberg.

In fact, agricultural work has been identified as one of the most dangerous occupations in the United States (Donham and Thelin 2006; McCurdy and Carroll 2000; McCurdy et al. 2003). For many years, pesticide exposure has been an area of increasing concern among farmworkers and their surrounding communities (figure 10.2). In 2000, 188 million pounds of pesticides were applied to crops in California (California Department of Pesticide Regulation 2020), whereas in 2006, 189.6 million pounds of pesticides were applied. Over a six-year period, the total amount of pesticides used increased by 1.6 million pounds. Pesticides have been shown to be related to rates of childhood cancer (Daniels et al. 1997) as well as an increased risk for neurological symptoms related to cumulative lifetime use of organophosphate and organochlorine (Kamel et al. 2007). In addition, children are at a much higher risk for developing problems associated with the exposure to pesticides than adults (Coronado et al. 2004; Etzel et al. 2003).

Figure 10.2. A photo of a warning sign that is required to be posted along the edges of agricultural fields when pesticides have been applied.

Photo courtesy of Steven J. Steinberg.

ISSUE BACKGROUND

Each year, millions of migrant and seasonal farmworkers are employed to work in the agriculture fields of the United States, many of them in California. In 2018, it was reported that agricultural workers held about 876,300 jobs and that the majority of farmworkers (over 68 percent) worked for crop and livestock producers (Bureau of Labor Statistics 2019). In terms of the job outlook, little or no change is expected in farmworker jobs during the decade from 2018–2028 (Bureau of Labor Statistics 2019). California's "agricultural abundance includes more than 400 commodities," making it both an important and an extremely diverse component of the state's economy (California Department of Food and Agriculture 2019).

California produces "over one third of the country's vegetables and two-thirds of the country's fruits and nuts" (California Department of Food and Agriculture 2019). While these statistics are impressive, they also underly the widespread use of agricultural pesticides in California's agricultural industry, which have the potential to pose a variety of health threats to agricultural farmworkers and those living in the surrounding communities. As a state, California employs the greatest number of farmworkers in the country (Bureau of Labor Statistics 2008). In fact, nine of the top ten agricultural counties nationwide are in California (United States Department of Agriculture 2019).

SOCIOSPATIAL GROUNDED THEORY: A COMMUNITY-ORIENTED APPROACH

This case study employed an approach we refer to as sociospatial grounded theory in combination with a mixed methods approach to examine community health issues related to pesticide use in rural California. Sociospatial analysis considers space, place, and social indicators in a holistic and integrated fashion (Steinberg and Steinberg 2009). An approach based in sociospatial grounded theory values the perspectives of community members as a linchpin of both problem definition and the subsequent research process (Steinberg and Steinberg 2006; 2015). This project particularly focused on the use of public participation research methods to capture local knowledge about pesticide use and the local community. There were three major components to the participatory aspects of this study: community organizers were interviewed to narrow down useful topics to focus on, organization members interacted through participatory geographic information systems (GIS) exercises and helped to ground truth statistical data, and finally, the results were analyzed and returned to the community organization for use in directing policy change. Primary data was paired with existing, secondary data about the community demographics, agricultural production, and pesticide use for the communities included in the study.

The research we conducted focused on promoting authentic community-based knowledge using an interdisciplinary mixed methods approach that included GIS and both qualitative and quantitative data collection methods to effectively integrate and triangulate environmental data, pesticide use information, and social data relating to community members' health. Applying methods rooted in sociospatial grounded theory is valuable because it originates in the field beginning with the local people's observations of problems or issues of importance to their community (Steinberg and Steinberg 2006; 2015). In this study, this approach provided an effective way to highlight the intersections between environmental and social issues for the farmworkers.

Our project used GIS to effectively integrate and overlap environmental and social data related to farmworker health and pesticide use. In this case, GIS was used as a data collecting tool, using an approach known as public participation GIS (PPGIS), as well as an analytical tool to compile and assess data. PPGIS is a technique for obtaining georeferenced data derived from local knowledge. The study drew on both important local knowledge from within the community and quantitative GIS and tabular data obtained from state databases. The sociospatial approach was especially valuable in this study because it allowed for the spatial portrayal of social and environmental data in a manner that was familiar to community members. In this case, the sociospatial approach highlighted the interplay between environmental and social issues for farmworkers in the region.

A role for qualitative and quantitative

It is common for research focusing on the impact of pesticide use to be strongly rooted in quantitative research. The role of GIS is key to examining spatial relationships, such as where pesticides are applied and in what quantities, alongside demographic data about the communities and populations potentially exposed (Gunier et al. 2001; McCauley et al. 2001). Some studies also include quantitative analysis of surveys to analyze possible routes of exposure to pesticides, noting that many routes of exposure also put members within a worker's home at risk (McCauley et al. 2001; Gomes et al. 1999; Gladen et al. 1998).

Qualitative studies have also focused on this topic, often showing a disparity between farmworkers and farmer owners in how they perceive pesticide exposure and risk (Arcury et al. 2001; Rao et al. 2006). Current qualitative research focuses on the need for education of farmworkers' families regarding risk-mitigating behaviors (Goldman et al. 2004; Rao et al. 2006). Differences in perceptions of pesticides suggest the need for a grassroots approach to pesticide education while taking care to address cultural differences in educational methods.

Using a sociospatial approach comes into play here because it allows for the spatial data collection and analysis of social and environmental data that relies on both qualitative and quantitative data in ways that actively engage with the community. In this case, our methodological approach started on the ground, beginning with the local people's observations of problems and issues that they believed to be important to their community. It also highlights the interplay between environmental and social issues for farmworkers in the region. We find that the visual approach to data portrayal so effectively supported through maps and GIS can effectively remove barriers that arise in studies working across a wide range of education, language, and literacy levels. This is a wonderful way to engage with a community and to draw on local knowledge and lived experiences.

Value of the mixed methods research

Our study used a bottom-up approach, beginning with networks of farmworker organizations, in order to intimately capture the concerns of persons working directly with pesticides or those who work in fields where pesticides are applied (figure 10.3). In other words, the beginnings of the project originated with the local community, building on their own observations and concerns about local community health and pesticides. However, farmworkers face a significant risk of

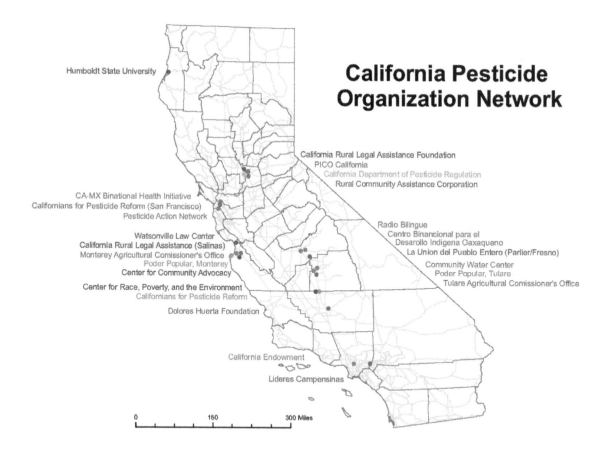

Figure 10.3. A map of key organizations working on pesticide issues in California at the time of the study.

Map courtesy of Steven Steinberg and Sheila Steinberg.

retribution from their employers if they raise questions about local working conditions. Farmworkers without documentation are at risk of firing and deportation with few legitimate avenues for complaint. Using a qualitative backbone in the initial construction of the project, we could better ensure that perspectives often unspoken by farmworkers would be brought to the forefront and highlighted. As a population that is used to being swept aside or ignored, they were given a chance not only to be heard but for their knowledge and experience to be truly valued and respected, perhaps for the first time in their lives. Often those conducting research on pesticides do not seek out this on-ground perspective, and because of some of the issues raised, including honest thoughts and opinions, can be risky for both the researchers and the participants.

Using GIS, we were able combine the qualitative stories of the farmworkers' experiences with quantitative data that highlights the types, amounts, and locations of pesticides. This mixed methods approach combines both theoretical and practical disciplines, such as sociology and ecology. This approach ultimately produces a more holistic view of a given problem (Greene 2007). Another benefit of using GIS in this study is that it facilitated the examination of both the quantitative and the qualitative mapped data while also providing a means to aggregate and anonymize the information we collected within the community. The resulting quantitative and qualitative data could be overlaid in GIS to identify patterns and trends between and among them, which was particularly useful when viewing both types of data simultaneously during analysis (Steinberg and Steinberg 2006).

This offers a path by which theoretical and practical information can be visualized in a format accessible by the layperson, making it particularly useful in applied studies in policy. This methodology is also useful for purposes of triangulation (Tashakkori and Teddlie 1998; 2003). Triangulation, a concept originating from mapping but used by social scientists, is a natural fit for GIS due to the ability to layer data on top of one another to allow for spatial analysis (Steinberg and Steinberg 2006). This solution calls for a specific, yet still relatively novel, mixed methods design: sociospatial grounded theory.

Local community networks and groups

One can never overestimate the role of community groups and networks in empowering members of a given community. At our six research sites, community-based empowerment groups were already working with the local community. One in particular, Poder Popular, was a longstanding and well-respected community-based organization with the trust of and a strong connection to the farmworker community. Poder Popular was a participating organization within the California Endowment's Agricultural Workers Health Initiative (AWHI). It is a community empowerment group seeking to support and improve the health of agricultural workers and their families through community organizing. At the time we were invited to conduct this study, the organization was seeking more detailed information about the effects of pesticides on those working in the fields of Monterey and Tulare Counties in California. Since our research work was supported by the California Endowment, we had direct access to Poder Popular and, by extension, the trust of the local communities. Poder Popular works in conjunction with a number of governmental agencies and other grassroots organizing groups to further their collective objectives of community health and well-being by leveraging existing resources in community. Their stated goals included a focus on population health and systems change, including the policies and protocols relating directly to pesticides in these communities where agricultural workers and their families are the most likely to be impacted.

Public participation GIS

The field portion of the study was conducted over a period of about six months involving a variety of key informant interviews, community meetings at which we conducted PPGIS activities, and secondary data collection, aggregation, and analysis. PPGIS is a process during which individuals participating and ultimately affected by the research provide feedback on aspects of the topic of interest, largely by responding to questions and discussion of their community experience by drawing on and annotating maps of the community.

The community of interest, which in this case is the farmworkers and agricultural community members, interact with mapped information and the researchers in order to produce relevant data based on their personal on-the-ground experiences. This serves not only as a vector for unique qualitative data but also as a method of triangulation for previously collected quantitative data. After processing and analysis, the data is then returned to the community for their own use along with the results of the study. While PPGIS may be accomplished by bringing computers with GIS software and data to the field, we opted for a low-tech solution to ensure accessibility to all community members. Bringing complex technology to the field may be perceived as intimidating to the community of interest. Instead, we brought large scale, printed maps and colored markers to the field to facilitate the PPGIS process. Participants in the PPGIS activities were then asked to draw and annotate their

Figure 10.4. Community members annotating large-scale community maps in a public participation GIS session. Different colors were used to indicate information representing various topics of interest such as pesticide incidents, at-risk locations when children and families spend time outdoors, and other key data.

Photo courtesy of Steven Steinberg.

observations and experiences directly on the maps, which we later captured into a geospatial database (figure 10.4).

The goal of the various PPGIS meetings was to allow members of the community an opportunity to document their local knowledge and experience, including issues that may not otherwise appear in the formal record. The process also engages the community in a fun, interactive activity that sparks conversation and eases apprehension, particularly for those who may feel intimidated or lack education and language skills, which leads them to avoid participating in very structured, formal surveys or interviews. Using a PPGIS approach helped us to gain valuable feedback from community members and also provided us an opportunity to validate secondary data sources that we had assembled through that point in the project. There are two main objectives to the PPGIS exercise: (1) ground truthing existing data for accuracy and (2) harnessing and capture local indigenous knowledge about relevant topics and information.

Throughout the PPGIS process, research assistants made observational notes, assisted with Spanish translation and map orientation for the participatory mapping exercise, and talked with participants about their experiences with farmworkers and pesticides in the region.

Meeting attendees were asked the following questions:

1. In your opinion, were mapped items accurate?

2. Where are the schools located?

3. Which schools did farmworkers' children attend? Which do nonfarmworkers' children attend?

4. Where are the farmworker, Latino, and white communities located?

5. Are there any notable locations of pesticide drift, spills, or other incidents related to pesticide use?

6. What type of crops are grown in specific fields?

In addition to the community PPGIS sessions held in each of the six study cities, three additional meetings were organized with representatives from Poder Popular and other interested community members. The research team presented GIS-based pesticide maps of the Monterey County communities of Salinas, Gonzales, and Greenfield and engaged in PPGIS methods with the meeting attendees. Attendance totaled about 20 in the city of Monterey (Monterey County), 12 at the meeting in Fresno (Fresno County), and 40 at a meeting in Visalia (Tulare County).

The use of PPGIS in all of these contexts helped to focus and validate our study objectives. This process is so valuable because it draws out the knowledge, opinions, and participation of those local to the community by using visuals, maps, and small-group discussion in conjunction with GIS technologies (Steinberg and Steinberg 2006). By interacting with the maps, community members helped us to produce and document relevant knowledge in a spatial context, offering a deeper understanding of ground conditions than would be possible using other methods. Leveraging the existing connections and community trust brought by Poder Popular, we were able to gather together diverse individuals who represented the local community, including farmworkers with in-depth on-the-ground experience and knowledge. The issue with the pesticides is that they affect the entire community, not just those in the fields, so drawing on the breadth of knowledge of community members provided insights that may have otherwise gone overlooked.

SOCIOSPATIAL GROUNDED THEORY AS MIXED METHOD DESIGN

Sociospatial grounded theory is an extension of grounded theory, defined by Glaser as a method "based on the systematic generating of theory from data that itself is systematically obtained from social research" (1978, 2). During traditional grounded theory research, the goal of analysis is to reach a point of "theoretical saturation" (Dey 1999). Sociospatial grounded theory extends these concepts to include the use of geographic statistical data (Steinberg and Steinberg 2006; 2015). The steps for sociospatial grounded theory begin with determining a topic of interest followed by a geographic location of interest. Then researchers collect the data for the research with linked spatial locations. Data is geocoded to provide a real-world spatial location for incorporation into the analysis, and ultimately, generation of both spatial and social theory occurs (Steinberg and Steinberg 2006).

This study used a mixed methods approach, including extraction of secondary pesticide data from the California Department of Pesticide Regulation database (2005); acquisition of geospatial data from a variety of local, state, and federal agency sources; key-informant interviews; PPGIS; and fieldwork including windshield reconnaissance and ethnographic methods.

Interviews

Interviews were conducted with leadership from the Latino community and people who were a part of the Poder Popular network. As the project progressed, we expanded our interviews to include others who were not a part of the initial network, including local, state, and federal government officials. Key-informant interviews were conducted in the first phase of the project for a strong understanding of issues the community faced related to health and pesticide use. Face-to-face interviews and phone interviews were conducted with a total of 16 key informants. Key informants consisted of community leaders, county officials, farmworkers, and Poder Popular leaders. The interview included 19 questions and took approximately 40 minutes to complete.

We used a snowball sampling method, a nonprobability sampling technique, to select key informants for interviews (Babbie 2013). This method involved beginning interviews with members of the target population and concluding the interview by asking them to recommend other possible informants. The term *snowball* is used because it is "a process of accumulation as each located subject suggests other subjects"

(Babbie 2013, 180). Using this technique, eventually the researcher begins to see the names suggested for additional interviews to repeat the same names numerous times, at which point a saturated sample has been achieved. In our work, this was particularly useful since the community organizations and key participants operated within a dense social network, strongly linked to other organizations working toward similar goals; these organizations also had a wealth of intimate knowledge concerning the work surrounding pesticide use.

A snowball sampling method was also useful given the time constraints of this study and the sensitive nature of the topic. Our original list of informants came from a list of partners related to the AWHI Poder Popular network. Additionally, we organized interviews with county agricultural officials in both Tulare and Monterey counties. Our research team established contact with potential informants using both email and telephone. Our research focused on the following general questions: Where were pesticides of concern located? Who were the local populations in close proximity to pesticides being sprayed and how were they affected by these applications? What would the implementation of a buffer zone around school property entail?

Mixed methods use

Study sites

During interviews, the research focused on six Central California towns within two Californian counties, Monterey and Tulare. The towns were Greenfield, Gonzales, and Salinas in Monterey County (figure 10.5); and Lindsay, Woodlake, Cutler, and Orosi in Tulare County (figure 10.6). An interesting local cultural phenomenon is how local people refer to the separate towns of Cutler and Orosi as if they were a single place, Cutler-Orosi, given their adjacency and sharing of goods and services between the two towns.

Pesticide application data was obtained from the California Department of Pesticide Regulation's 2005 statewide pesticide database. This database provides detailed information regarding pesticide applications for each public land survey section in the state reported in pounds of each individual active ingredient. This database was linked to a geospatial database for each study county using ArcGIS. Additional data obtained included county parcels and zoning information and the locations of all schools for each of the six cities analyzed in the study. California State Senate and

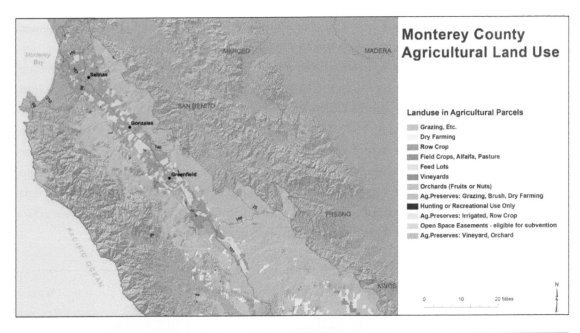

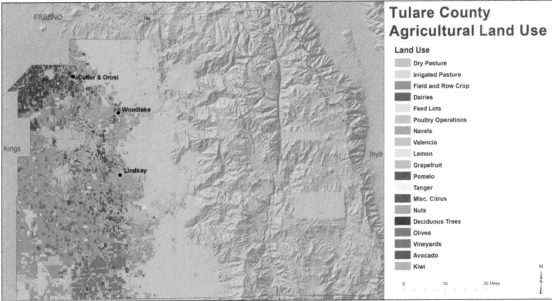

Figures 10.5 and 10.6. The Monterey County study region (*top*) included the three focal communities of Salinas, Gonzalez, and Greenfield along US Highway 101 running from northwest to southeast in the Salinas Valley. The Tulare County study region (*bottom*) included the three focal communities of Cutler and Orosi, Woodlake, and Lindsay from northwest to southeast along the southwestern edge of Sequoia National Forest.

Maps courtesy of Steven and Sheila Steinberg.

Assembly districts were also mapped to identify which elected officials represented regions where the highest levels of use of agricultural chemicals were noted.

Environmental data included topography derived from the 10-meter resolution National Elevation Dataset (NED). To examine winds, we obtained wind hindcast data from the National Oceanographic and Atmospheric Administration (NOAA) for the week of July 1–7, 2006. This data included hourly wind speed and direction hindcasts at 5-square-meter resolution. Wind data was used to develop maps showing a weekly average wind speed and direction at four-hour intervals for a representative week of the growing season when pesticides are typically applied. Wind data was preprocessed using NDFD GRIB2 software (National Weather Service 2007) and exported as shapefiles for use in the GIS analysis (figure 10.7).

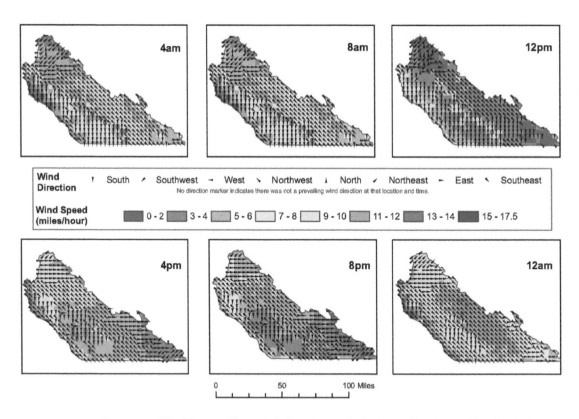

Figure 10.7. Weekly prevailing wind direction and wind speed by time of day for July 1–7, 2006, in Monterey County, California. Speeds were averaged for four-hour blocks over a 24-hour period. Arrows indicate the prevailing wind direction in 45-degree intervals. The colors indicate wind speed in miles per hour. Strongest winds occurred during the overnight and early morning hours.

Map courtesy of Steven and Sheila Steinberg. Wind data from the National Weather Service.

Sociospatial methods

The various sociospatial methods described previously were integrated into a methodological workflow beginning with key-informant interviews and ethnographic methods. This was followed by baseline data compilation of maps and data for the PPGIS sessions. We overlaid National Agriculture Imagery Program (NAIP) color images of the six study communities with GIS data showing roads and parcels as a base for community members to provide input. Prior to field visits, we interviewed key informants (targeted individuals from government agencies, community groups, and nonprofit organizations working on pesticide issues) to determine the issues and active chemical ingredients of greatest concern. Field visits to both counties were made in the summer of 2006 to carry out a community-based participatory research (CBPR) approach as a means to understand community members' interests and knowledge about pesticides. Particular emphasis was placed on the amount and types of specific pesticides and fumigants used near schools, neighborhoods, and community gathering places. Participants mapped locations where they had observed or experienced pesticide drift. The PPGIS process was also used to identify places where community members participate in outdoor activities such as parks and open spaces, as well as locations where farmworkers live. The PPGIS approach provided an additional and extremely rich dataset based on local knowledge and was useful to validate existing geospatial data sources (for example, United States Census Data and the respective county assessor's office) by cross-referencing the PPGIS data with these sources as a means of accuracy assessment (figure 10.8).

Local voices and quotes

Throughout our research process, a series of qualitative key-informant interviews were conducted. This section presents examples of the stories gathered from farmworkers and community members as we sought to better understand the lived experiences of the community and issues they faced. These semi-structured interviews provided ample opportunity for participants to share their stories and impressions. The key-informant interviews and participatory mapping exercises yielded a wide variety of qualitative data in the form of stories, comments, and written notations relating to issues of pesticide and health. Through our research, it became clear that community members had an awareness that unusual health issues were experienced by their communities, but they were not sure how to put their finger on what it was or how to safely communicate these concerns or take action to address issues they believed needed to be addressed.

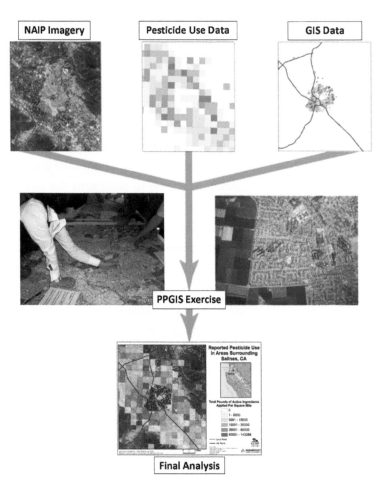

Figure 10.8. PPGIS workflow used in this project. Data from multiple geospatial sources, including aerial imagery, existing GIS data, and modeled GIS data, were brought together to produce large-format maps of each study community. Community members annotated maps through a public participation mapping exercise to provide information about sensitive sites, areas where pesticide drift occurred, and other relevant information relating to the analysis.

Figure courtesy of Steven and Sheila Steinberg.

UNCERTAINTY, HEALTH, AND PESTICIDES
Pesticides and communities

Another interesting theme that emerged from our interviews and PPGIS sessions was the notion that pesticides impact everyone in the community regardless of if they are a farmworker or not. It was the big theme that pesticides don't discriminate at all, and they drift to wherever the wind is blowing, including into neighborhoods or schoolyards adjacent to the fields (figure 10.9).

Figure 10.9. A school playground sits immediately adjacent to wide-open fields with no protection or buffer of any significance. The distance from the school fence to the crop is slightly wider than a single-track dirt road along the perimeter of the field.

Photo courtesy of Steven Steinberg.

As one interviewee noted:

> Even if they aren't working in the fields, they are affected by the fields because the town is surrounded by the fields. Everybody fits the guise of being a farmworker because you are getting what is going on in the field. The communities are getting 95 percent of the crap that comes off of the fields.

The point that everyone is impacted by the spraying of pesticides in a community is a common theme that emerged.

Another interviewee commented:

> People get exposed when they are driving around and when they are at school. It's impossible to pick one story. We have strong anecdotal evidence. People are getting drifted upon all the time.

These shared quotes from local place-based interviews and stories highlight that people are affected by the pesticide spraying as part of the agricultural industry, which so many of them participate in. But it's not by choice that other members of the community are being exposed elsewhere in the agricultural communities.

Clearly, people are unsure about the impacts of pesticides, and they have a keen fear associated with it. One interviewee shared:

> Specific to pesticide drift, in my opinion, there is not a strong understanding of the chemicals, which ones are dangerous. What are the maximum drift potentials? How far will it go, windy days versus nonwindy days? What is lasting drift potential and where will it go?

On the topic of community health and pesticides, we asked, *"How do people know when pesticides are being used in their community?"* An interviewee shared the following observation:

> You might not know. If you smell something different or if you see the airplanes, driving through you see it, but sometimes you don't. They might have already sprayed ahead of the time you go through it.

Another interviewee noted:

> How many people are hurt? We don't know; we don't know those people who are hurt immediately. ... Pesticides have been related to much more long-term effects, such as Parkinson's, infertility, etc. We need a breakthrough that can only come through research.

Another interviewee commented:

> My daughter is 17 now and has severe migraine headaches. We all got really
> sick. My grandson is autistic; I think there is a connection between autism
> and pesticides. My other grandchild had a rash for three years, and it finally
> went away at age five. They were applying fumigant agents, such as Metam
> sodium and Iodomethane.

Clearly from this quote, the community member is aware that something is amiss
in their neighborhood, but they don't know exactly how to put their finger on it.
While these observations are anecdotal, they were consistent across most interviews
and indicative of a larger narrative that people are experiencing health maladies as
part of the community where they live and work. Many of the people we spoke with
were fearful to share too much information for fear of losing their jobs or coming
into conflict with local authorities.

As shared earlier in this chapter, there are many risks associated with agricultural work and the use of pesticides. Accidently getting sprayed by them or spray
affecting local communities are two such risks. As part of the storytelling process,
we found that farmworkers often have no idea of the dangerous field situation that
they might be walking into, but they must show up to work and be ready to do their
job (figure 10.10).

Figure 10.10. Farmworkers in the fields are sometimes unknowingly exposed to recently
sprayed pesticides, or overspray and drift may enter into the neighborhoods immediately
adjacent to the fields.

Photo courtesy of Steven Steinberg.

For example, one woman whose husband is a farmworker shared, "My husband was working in the fields and the contractors did not advise the supervisor and the group of farmworkers that they had sprayed the day before. The contractor wanted to finish that field, so he sent people in to work. My husband developed an allergic reaction and headaches. He ended up applying to work for the factory."

Being able to share an account such as this highlights that people feel a lack of decision-making power for the situations that they face on a daily basis. So, they are, in effect, powerless to walk away from their economic livelihoods and have to continue to work in these jobs to support their families.

On the topic of safety, one interviewee shared the following:

One thing we look to add is pesticide safety measures. Specifically, for example, right now under federal law, agriculture employers and commercial applicators are not required to keep records of all the pesticides they apply. Say, someone gets hurt, maybe not right away, like a birth with birth defects, and you want to know what they've been exposed to. There are not records to determine what the exposure was. And therefore hinders.

Oftentimes there is a clear lack of knowledge about exposure. One woman noted:

There is no notification for pesticides in California with the exception of schools and daycares. Those are the only places that require pesticide implementation. There is a strong noncompliance despite that it is a law. People only know when they are affected when they poison themselves or when they sense exposure. More often than not, people are being exposed and they don't even realize it. If they know they were exposed, they don't know what to do with it.

In the past, those in power, the growers who lead the agricultural operations that use the pesticides, have created barriers to putting in place effective safety measures. One interviewee explained:

Buffers around schools. Expand protection zone around schools. School boards have to vote on the pesticide buffer zones. ... Community members speak their cases to committees. ... A few cases have been successful, but three of the five board members were growers who didn't allow the issues to come up.

When it comes to pesticide use, there is a general question about when and how the pesticides are applied. People have reported that they know the pesticides are being used through one of their senses: sight, smell, hearing. One of the informants that we interviewed shared the following quote regarding nighttime application of pesticides, a topic that came up on multiple occasions and which we personally observed during our field research. Pesticides were commonly being sprayed in the middle of the night, which happens to coincide with when the wind speeds are the highest (figure 10.7). One interviewee said:

> Sometimes they [farmworkers] just see an airplane or helicopter or tractor. But a lot of the time, these are done in the middle of the night. Not so much to hide them, but they don't want to do them while people are in the fields.

At first it might seem like a good idea to spray pesticides in the middle of the night when workers are not out in the fields. However, on further investigation, we discovered that there is a significant downside to nighttime spraying as well, which is a potential lack of awareness that spraying has occurred at all. When kids go to school the next day and pesticide drifted onto the playground equipment, there may be no way to know about the risk until it's too late and kids are experiencing negative health reactions.

Other times, it's the sense of smell that might alert farmworkers to the recent spraying of pesticides. One interviewee shared:

> We smell it. The ones that know about it will come and let me know so I'll report it, and others don't know to report it, so they just accept it.

Another person shared:

> Typically, they see the crop duster, or the folks doing it on the fields smell it at times, and sometimes on the fruit they see when they pick it [figure 10.11]. It's really bad in the cotton industry; they really have to cover up. They have nasal problems.

Figure 10.11. Farmworkers picking crops in the fields are assessed by the amount they pick and box in a shift. There is little time to consider what unknown risks may be present relative to pesticide exposure.

Photo courtesy of Steven Steinberg.

Regarding health and the actual diagnosis of being exposed to pesticide use, sometimes there are sketchy processes in place to diagnose problems associated with potential pesticide exposure.

SOLUTIONS

It's clear that people in these agricultural communities recognize there is a problem associated with pesticide drift and that something should be done about it. The quote from an interviewee below highlights the need for people to have jobs in the agricultural industry and to have a safe working environment as well that does not negatively impact their health.

> The strategic work around the buffer zones and the community health piece is a major concern. … What we need to be doing is working toward real solutions for what it takes to support growers to use different methods of growing. This push and pull method doesn't work. We need to figure out together how society can support community health while keeping the prosperity of the agriculture.

Gaining insights

These interviews provided insight into primary issues of relevance related to pesticide use, pesticide drift, agricultural worker health, and community issues for these regions. Analysis of these interviews helped us to understand issues related to pesticide drift, how farmworkers are affected by drift, where and how they go to seek treatment in instances of possible exposure to pesticides, and other specific areas of concern in these communities. Furthermore, by conducting interviews with various parties related to this issue including nonprofit groups, agricultural and county officials, farmworkers, and community members, issues of importance for policy emerge. The interviews provide the basis of the farmworker/pesticide drift interaction and perspective presented in the final study report (Steinberg and Steinberg 2008a).

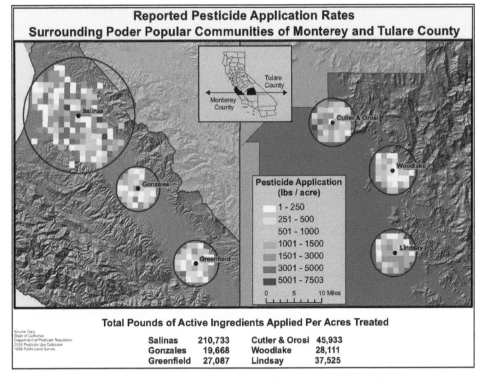

Figure 10.12. The total amount of pesticide applied in the area surrounding each of the six study communities in pounds per acre of field. Pesticides are aggregated to the public land survey grid, with cells of approximately one square mile (640 acres).

Map courtesy of Steven and Sheila Steinberg. Data on application rates comes from the California Department of Pesticide Regulation.

Pesticide data was examined for the region surrounding the three Poder Popular communities in each of the study counties. We examined the total amount of pesticide applied in 2005 as well as the amount of individual pesticides of concern. Figure 10.12 shows an example map for the distribution of pesticide application in pounds per acre of active pesticides in the areas surrounding each of the study cities. The Department of Pesticide Regulation summarizes application rates by public land survey sections of approximately one square mile (640 acres). As an approximation of the total pesticide impact associated with a particular community, we examined a region approximately three times the diameter of the developed area of the community. For the five smaller communities of Gonzales, Greenfield, Cutler and Orosi, Woodlake, and Lindsay, the developed area of the community was approximately one mile in diameter. Thus, we totaled all pesticide applications for 2005 within a three-mile radius of the center of the community (approximately 28.3 square miles).

The city of Salinas in Monterey County is substantially larger in size, measuring approximately three miles in diameter, so for Salinas, we totaled all pesticide applications for 2005 within a nine-mile radius of the center of the city (approximately 254.5 square miles). Salinas has a unique layout resembling a doughnut with a significant area of agricultural production remaining in what is now the middle of the developed area of the city and located across the street from a major hospital. Table 10.1 summarizes the total pounds of active ingredient and average pounds per square mile applied in and around these communities. Rates of pesticide use were relatively higher for communities in Tulare County.

Table 10.1. Amount of active ingredients applied around each study community in 2005 reported in total pounds and average pounds of per square mile

County	Community	Total pounds of active ingredients applied in 2005 (area considered)	Average pounds of active ingredients applied in 2005 (per square mile)
Monterey	Salinas	210,733 (254.5 mi²)	828
	Gonzales	19,668 (28.3 mi²)	695
	Greenfield	27,087 (28.3 mi²)	957
Tulare	Cutler & Orosi	45,933 (28.3 mi²)	1623
	Woodlake	28,111 (28.3 mi²)	993
	Lindsay	37,525 (28.3 mi²)	1326

A topic of particular interest among community members interviewed in this study was the establishment of buffer zones around schools. A buffer zone is a region of some specified distance around school grounds with limitations on how and when particular pesticides may be applied. At the present time, the creation of buffer zones and the size and other regulations are determined on a county-by-county basis. To assist in illustrating the impacts of a proposed buffer zone of a quarter mile (1,320 feet), the maximum proposed at the time of this study, we identified schools for the entire county as well as for each of the study communities. Assessor's parcels associated with school structures were overlaid on the NAIP photography and inspected to ensure all playgrounds and athletic fields associated with the school were included in the analysis. In many cases, a school owned multiple parcels, and buffers had to be generated for the total extent of the grounds rather than just the location of the school building.

For each study community, we aggregated the total acres of agriculturally zoned land falling into the quarter-mile buffer zones as well as the percentage of the total amount of agricultural lands surrounding the community. For areas surrounding towns, this naturally overestimates the percent of agricultural lands affected relative

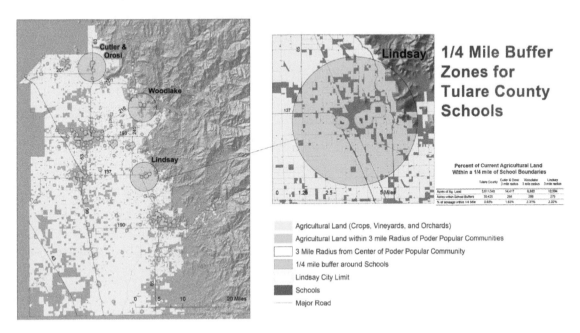

Figure 10.13. An examination of the impact that a hypothetical quarter-mile buffer zone around schools would have on agriculturally zoned lands in Tulare County, California.

Map courtesy of Steven and Sheila Steinberg.

to the outlying areas of the county. For example, in Tulare County, of the over 5.6 million acres of agricultural land, 35,420 acres were located within a quarter mile of a school or school grounds, representing 0.63 percent of the total (figure 10.13). In Tulare County, within the three-mile radius (28.3 square miles) surrounding each of the Poder Popular communities, buffer zones would have the largest impact on agricultural lands in the community of Woodlake with just over 2.3 percent (208 acres) of agriculturally zoned lands falling within a quarter mile of a school. In Lindsay, just over 2.2 percent of these lands are within a quarter mile of a school and for Cutler and Orosi, just over 1.8 percent.

Exploring this further, it is apparent that many schools are located at the periphery of these and other communities, perhaps due to the availability and price of this land as urban and suburban development grow outward from the city center. We observed very similar patterns and results in Monterey County (figure 10.14).

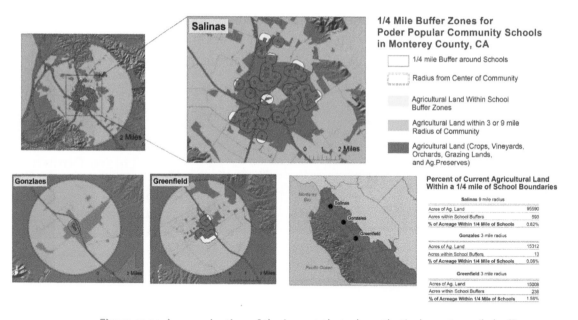

Figure 10.14. An examination of the impact that a hypothetical quarter-mile buffer zone around schools would have on agriculturally zoned lands in Monterey County, California.

Map courtesy of Steven and Sheila Steinberg.

CONCLUSION

This chapter explored the connection between health and environment for a variety of agricultural communities in two counties in California. The lessons learned through this process of using multiple research methods and specifically PPGIS are that local people possess valuable knowledge about the social, health, and environmental conditions where they live, work, and play.

Our research also highlighted how there are times when gaps exist in the scale of publicly available data, but the missing data can be contextually completed through engaging with local populations and drawing on their knowledge. While no pesticide data was available at a finer scale than one square mile, participants were often able to identify streets or fields related to incidents of experienced pesticide drift. Cross-referenced demographic data confirmed the validity of differences perceived by participants; however, participants were often able to identify demographic regions at a much finer scale than the US Census block group level. Sensitive sites within the community are typically not included in land use or parcel data, as some community locations, like churches or residential homes, had secondary purposes unique to the community served and would have otherwise gone unnoticed as sensitive.

We can improve research for such types of projects in the future through increasing our awareness of community perceptions of space. Participants were largely unfamiliar with aerial imagery and map-reading and sometimes required assistance finding locations on the maps produced for the exercises. However, community members found location less difficult when the main local roads through town were labeled, despite the labeling of freeways and interstates. It would have been useful to include questions regarding local landmarks and heavily utilized streets in the interviews for labeling maps prior to beginning PPGIS exercises. Including a number of insets displaying the study area at varying scales without the background confusion of aerial imagery may have also increased cartographic orientation on the part of community members.

One final challenge facing researchers is obtaining community-based knowledge in the face of widely available online mapping programs. During one exercise, researchers observed one participant using a laptop to confirm the locations of schools and obtain their listed names, labeling these on the map. Care should be taken to remind community members that the goal of the process is obtaining information specific to their experiences as opposed to data already publicly available.

FINAL THOUGHTS

Since the time that we completed and reported on this work back in 2008, local community members have successfully argued for pesticide buffer zones and prior warnings before pesticides are sprayed around schools, especially in Tulare County. It's interesting to note that it took a great deal of time and effort for local people's voices to be heard and acknowledged in the form of policy. But as we had anticipated, the resilience of the community comes from within, and using the integrative capabilities of GIS, in concert with a mixed methods research approach, the effort paid off. Policy and change do not always come quickly, but as is so often the case, by working together, sharing their stories, and ultimately documenting them with the power of maps and data, the community found success in improving their situation despite the fact that it took longer than one might hope for this to come to fruition. A *story* (earlier known as a *story map*) (Steinberg et al. 2017) provides additional details of the larger issues and outcomes.

As researchers who engage in community-based participatory research, we believe it is important to underscore the importance of community engagement when conducting research in any community. While it's common for researchers to come into a community with preconceived notions of the problems to be studied and solved, the results are much more valuable when you begin by asking local individuals about their experiences, desires, thoughts and ideas about ways to improved their own community and build lasting resilience.

ACKNOWLEDGMENTS

This project was funded by The California Endowment Agricultural Workers Health Initiative. We wish to thank the California Endowment for support of this project. We would especially like to thank Mario Gutierrez, M.P.H., director of the Rural and Agricultural Worker Health Program, for his vision in supporting this work.

We would like to thank the students who assisted on this project: Jennifer Kauffman, Nanette Yandell, Josef Eckert, Marian Strong, Launa Peeters, and Adrianna Bayer. Their contributions were invaluable to the success of this project, and for each the opportunity to participate in this important project and to meet so many wonderful people was both enlightening and life-changing in many ways.

REFERENCES

Arcury, T. A., S. A. Quandt, A. J. Cravey, R. C. Elmore, and G. B. Russell. 2001. "Farmworker Reports of Pesticide Safety and Sanitation in the Work Environment." *American Journal of Industrial Medicine* 39: 487–98.

Babbie, E. 2013. *The Practice of Social Research, 13th Edition.* Belmont, CA: Cengage Learning.

Bureau of Labor Statistics. 2019. Accessed on December 29, 2019. www.bls.gov.

California Department of Food and Agriculture. 2019. "California Agricultural Production Statistics." Accessed on December 20, 2019. www.cdfa.ca.gov/statistics.

California Department of Pesticide Regulation. 2020. Accessed on December 20, 2019. www.cdpr.ca.gov.

Coronado, G. D., B. Thompson, L. Strong, W. C. Griffith, and I. Islas. 2004. "Agriculture Task and Exposure to Organophosphate Pesticides among Farmworkers." *Environmental Health Perspectives.* 112: 142–47.

Daniels, J. L., A. F. Olshan, and D. A. Savitz. 1997. "Pesticides and Childhood Cancers." *Environmental Health Perspectives.* 105: 1068–1077.

Dey, I. 1999. *Grounding Grounded Theory.* Bingley, England: Emerald Publishing.

Donhan, M. and A. Thelin. 2006. *Agricultural Medicine: Occupational and Environmental Health for the Health Professions.* Ames, IA: Blackwell Publishing.

Etzel, R., S. J. Balk, J. R. Reigart, P. J. Landrigan. 2003. "Environmental Health for Practicing Pediatricians." *Indian Pediatrics*, September. 40 (9): 853–860.

Gladen, B. C., D. P. Sandler, S. H. Zahm, F. Kamel, A. S. Rowland, and M. C. R. Alavanja. 1998. "Exposure Opportunities of Families of Farmer Pesticide Applicators." *American Journal of Industrial Medicine.* 34: 581–587.

Glaser, B. 1978. *Discovery of Grounded Theory: Strategies for Qualitative Research.* Philadelphia, PA: Routledge Press.

Greene, J. C. 2007. *Mixed Methods in Social Inquiry.* San Francisco, CA: Jossey-Bass Inc.

Goldman, L., B. Eskenazi, A. Bradman, and N. P. Jewel. 2004. "Risk Behaviors for Pesticide Exposure among Pregnant Women Living in Farmworker Households in Salinas, California." *American Journal of Industrial Medicine.* 45: 491–499.

Gomes, J., O. L. Lloyd, and D. M. Revitt. 1999. "The Influence of Personal Protection, Environmental Hygiene and Exposure to Pesticides on the Health of Immigrant Farm Workers in a Desert Country." *International Archives of Occupational and Environmental Health.* 72: 40–45.

Gunier, R. B., M. E. Harnly, P. Reynolds, A. Hertz, and J. Von Behren. 2001. "Agricultural Pesticide Use in California: Pesticide Prioritization, Use Densities and Population Distribution for a Childhood Cancer Study." *Environmental Health Perspectives.* 109: 1071–1078.

Kamel, F., L. S. Engel, B. C. Gladen, J. A. Hoppin, M. C. R. Alavanja, and D. P. Sandler. 2007. "Neurological Symptoms in Licensed Pesticide Applicators in the Agriculture Health Study." *Human & Experimental Toxicology.* 26: 243–250.

McCauley, L. A., M. R. Lasarev, G. Higgins, J. Rothlein, J. Muniz, C. Ebbert, and J. Phillips. 2001. "Work Characteristics and Pesticide Exposures among Migrant Agricultural Families: A Community-Based Research Approach." *Environmental Health Perspectives.* 109: 533–538.

McCurdy, S. A., and D. J. Carroll. 2000. "Agricultural Injury." *American Journal of Industrial Medicine.* Sept. 6, 2000.

McCurdy, S. A., S. J. Samuels, D. J. Carroll, J. J. Beaumont, and L. A. Morrin. 2003. "Agricultural Injury in California Migrant Hispanic Farm Workers." *American Journal of Industrial Medicine.* 44: 225–235.

National Weather Service. 2007. NDFD GRIB2 version 1.85. https://vlab.ncep.noaa.gov/web/mdl/degrib-for-ndfd.

Rao, P., A. L. Gentry, S. A. Quandt, S. W. Davis, B. M. Snively, and T. A. Arcury. 2006. "Pesticide Safety Behaviors in Latino Farmworker Family Households." *American Journal of Industrial Medicine.* 49: 271–280.

State of California. 2020. Accessed on May 5, 2020. www.ca.gov.

Steinberg, S. L., and S. J. Steinberg. 2006. *GIS for the Social Sciences.* Thousand Oaks, CA: Sage Publications.

Steinberg, S. L., and S. J. Steinberg. 2008a. *People, Place and Health: A Sociospatial Perspective of Agricultural Workers and Their Environment.* Humboldt State University, Arcata, CA. http://hdl.handle.net/2148/428.

Steinberg, S. L., and S. J. Steinberg. 2008b. *People, Place and Health: A Pesticide Atlas of Monterey County and Tulare County, California.* Humboldt State University, Arcata, CA. http://hdl.handle.net/2148/429.

Steinberg, S. L., and S. J. Steinberg. 2009. "A Sociospatial Approach to Globalization: Mapping Ecologies of Inequality." *Understanding the Global Environment,* edited by Samir Dasgupta, 99–117. Delhi, India: Pearson Longman. ISBN: 9788131717028.

Steinberg, S. L., and S. J. Steinberg. 2015. *GIS Research Methods: Incorporating Spatial Perspectives.* Redlands, CA: Esri Press.

Steinberg, S. L., S. J. Steinberg, and N. R. Malloy. 2017. *People, Place, and Health: A Sociospatial Story of Agricultural Workers and Their Environment.* www.arcgis.com/apps/Cascade/index.html?appid=1705ab268dd64f34ae3906eb35ee6ee0.

Tashakkori, A., and C. Teddlie. 1998. *Mixed Methodology: Combining Qualitative and Quantitative Approaches.* Thousand Oaks, CA: Sage Publications.

Tashakkori, A., and C. Teddlie. 2003. *Handbook of Mixed Methods Research.* Thousand Oaks, CA: Sage Publications.

United States Department of Agriculture. 2019. Accessed on November 20, 2019. www.usda.gov.

CHAPTER 11
Spatial resilience policy and action

**SHEILA LAKSHMI STEINBERG AND
STEVEN J. STEINBERG**

BEST PRACTICES FOR USING LOCAL KNOWLEDGE

In this final chapter, we highlight some of the lessons drawn from the case studies, along with a number of best practices for using spatial thinking and tools in assessing, planning for, and building resilience and policies to support it. One constant across the examples we have presented here is that local people possess knowledge central to the places where they reside. In other words, those people who live in a particular place have unique knowledge and understanding of their environment, its resources, and the community itself. It stands to reason that individuals who spend days traversing, working, and living in a given locality will develop a more in-depth understanding of it. Not only do local folks know their way around a neighborhood, local environment, or community, but they also tend to understand the history, norms, and balance of the resources in their environment.

When there are significant changes to an environment—changes in the climate, the migration patterns of local fish and wildlife, or other changes to what the community understands to be the historical norms—there are likely to be changes to the associated resources in that place as well. Local communities will be the first to notice changes in resources, and in places where these resources are essential to the survival of the community, they will be the first to be affected by those changes. For instance, as weather patterns shift, the flora and fauna of the region may similarly shift or perhaps be lost due to a lack of climate refugia or opportunities to migrate. A community's ability to plan for or adjust to changes that impact it is a cornerstone of resilience.

Changes modify the people-place equation

The people-place equation describes how people develop a reliance on their environment over time to provide a certain set of natural, physical, and cultural resources that define and support their community and culture. When changes alter these relationships, major shifts can happen, and people must adjust to persist. Communities that understand their local places from a sociospatial perspective are better equipped to break down and to solve their own problems.

Better decision-making

GIS can serve as a valuable tool to capture and to integrate knowledge of local people, places, and resources. Why does this matter? It matters because if you can visualize data clearly as portrayed in a real-time map or spatial dashboard, it will lead to opportunities to make decisions. A person trying to better understand a problem or situation must ask questions about location, quantity, nearness (proximity), and relationships to other problems or situations. Such things taken together help to account for gaps, strengths, and opportunities in the data. The decisions made are only as good as the data used to inform them, so by leveraging the capabilities of GIS technology, related information, and local knowledge, communities can make decisions about their own future resilience.

Visualizing and communicating data

Data is the key to understanding challenges faced by a particular community or place. Throughout the examples in this book, we've explored the power of community knowledge, ranging from that of Indigenous people, local community members, professional managers, and academic researchers. Regardless of the case examined, the importance of place-names and a historical perspective are emphasized. The how, when, and where of actionable decisions and policies are best achieved when based on a holistic view of people and their communities.

An important part of any communication is the pictorial part of it. Visual images have a defining impact on underscoring patterns in the data. We all know the cliché that a picture is worth a thousand words. When it comes to telling stories, the verbal or written narrative is important but so too are the images or visuals that accompany these stories. Stories about space and place have existed throughout history and have been handed down from one generation to another. Before written language, many cultures captured their stories using symbols and images and through oral traditions passed down through the generations. These served as important methods of communication and documentation.

Capturing these in conjunction with and in context of modern digital tools such as GIS, ArcGIS StoryMaps, and other interactive tools can provide a richer means of visualizing and sharing this history and using it in context of managing and maintaining resilience for the future. Today we are inundated with data and information virtually all the time, perhaps to an excess. What this means in terms of visualization is that it's important to consider the context of communication. Visualization that uses maps helps to establish a broader picture and a wider range of information than any other option.

Diverse stakeholders

Throughout each of these chapters, we have attempted to provide examples of the value of GIS and spatial analysis as a communication tool that can work across a variety of audiences. In any situation in which a community is deciding how to best address a problem, manage for sustainability and resilience, and adjust to a changing world, the stakeholders involved will be relatively diverse. Maps can provide a valuable tool for organizing and exploring this information and highlighting places, spaces, and resources that are important to a community, culture, or society. This was true in the past and it is true today.

The challenge becomes how to identify and engage with a diverse group of relevant stakeholders. You must begin by identifying the demographics of the community to account for the various ethnic, cultural, and potentially socioeconomic or political subgroups within a place.

THE IMPORTANCE OF PEOPLE AND PLACE IN DEVELOPING RESILIENCE

In considering what makes people in a place strong and resilient, there are many factors to consider. What skills and abilities are present within the local population? How integrated are they with one another? Do diverse elements of the community value and support one another? What is the nature of the strong and weak social ties within a community, and how effectively can they be called upon to act? In other words, how well do people in the community have each other's back when a crisis emerges? As authors of this book, we have been fortunate to have lived in an extremely diverse range of communities throughout our lives—rural, suburban, and urban both within the United States and abroad—and to have traveled to and worked in many more. Throughout these many personal experiences, it has been interesting to observe how people in diverse places really do have the best interest of their fellow community members at heart, and how they actively come together to help one another.

Local community sustainability

One clear pattern we've observed is that people living in rural and remote environments depend on one another, especially when a crisis occurs. There is a much more internal sense of community sustainability in which the local people do not have an expectation that help will come from the outside, at least not quickly. In these circumstances, they step up to assist one another. This pattern is also visible in urban environments, especially in communities where neighbors need to rely on one another on a regular basis to meet their daily needs. When economic or political barriers arise and disenfranchise or isolate such communities, the internal social networks and community cohesion may provide a level of resilience to the community as essential to their survival as the availability of wild fish and game may be to a rural community. The specific resources differ, but the general concepts and methods are similar.

In such situations—when people need the help of their community members, people tend to pitch in and help. Perhaps the biggest fault of the suburbs is the extreme over-independence of the families who live there. Often these are middle- or upper-class, economically viable families living in suburbs where they have sufficient resources, which provide them with mobility and capability to fend for themselves without depending on one another for assistance (or, with some events, perhaps they simply don't believe they will need help until it's too late). In many cases, individuals residing in these suburban, less socially connected locations have the financial resources to accomplish things themselves without help from their community. In such places, this does little to foster a notion of community and in many cases actually detracts from it.

WHAT MAKES A COMMUNITY RESILIENT?

One may ask, what makes a community resilient? Further, what sorts of policies can be created to help communities across the board achieve resilience, and how can the consideration of space and use of tools such as GIS help? Communities are most likely to reach their full potential by becoming aware of the local resources available (including both social and environmental) and then building on those strengths using spatial, place-based action that uses these local ecologies, patterns, and assets. Some key concepts that may help to identify what makes a space and place resilient include the following.

Place matters

Local geography matters, as do the cultural, social, and political structures of the place. Some places encompass communities that are collectively committed to their success and sustainability over time, while others are not. In other words, for many communities, folks will go out on a limb to help one another out, often due to a deep sense of connection to the history and story of the place. This may be derived from a long or multigenerational history in the place, a particular affinity for the geography and environment of the place, or other cultural or professional ties to the particular location. For example, it might be a community in which people have generations of history or that is similar to the one from which the early settlers to the region had originally come from, offering a similar environment, even in a place that was more recently settled. Over time, places also draw cultural meaning from the very stories that revolve around the place and the natural or built environments there: a sacred hot spring, a centuries-old church, or a fishing village and its long-developed connection to the sea and the lighthouse at the entry to the harbor.

Adaptability and action

The degree of adaptability of a particular people in their place will be determined, in large part, by that population's local knowledge and understanding of their environment along with a strong belief that their actions can make a difference. A society that can identify, assess, and react to changes in its environment will be more resilient. But they must first understand the changes that have or likely will occur. Then, through careful consideration, interpretation, and understanding of this data, they must arrive at some action to undertake as a response. Second, there must be a set of skills or capabilities that can be brought to bear by the community in managing and adapting meaningfully with the local environment, or in some circumstances, to recognize the best option is to seek an alternative environment better suited.

Community connectedness

The degree of cohesion or connectedness that members of a community feel will impact their willingness to want to help in any response to changes in an environment. Usually, the connections that people create with one another in a place are built around shared meaning, engagement, and valuing of their community. Being able to identify and map a community's connectedness (social capital and human capital) can help to draw out the strengths of a local community. Another way to describe this might be to examine the level of community spirit. The spirit that a community feels can impact their desire to want to make a difference in promoting the community going forward. By contrast, less socially connected places, for example, a suburban bedroom community with high levels of population turnover, may have a much lower level of cohesion or community spirit because individuals are less fully engaged with their community. Less socially connected communities will likely face far greater challenges in coming together around issues that are core to building an adaptable and resilient community.

Symbolic physical features

Not only are people connected to one another, but they are also connected to the physical features of their environment or place. Many local communities are built on the notion that people and place are enmeshed together. The connections that members of a community feel for a place stem from their experiences there and the sentiment that one feels for aspects of a physical environment. Communities are made up of physical, social, and cultural resources. For example, there may be landmarks or other physical, symbolic aspects of a community that people orient to. Such symbols may come to represent the local people and place. Once a cultural symbol is identified by the local people, it comes to represent the place.

For instance, in the region near the towns of Silver City and Bayard, New Mexico, the residents had a strong connection to a rock formation known as the Kneeling Nun. The rock, perched on a nearby mountainside, looked much like a silhouette of a nun overlooking the mine that served as one of the primary employers for the community. People revered this rock formation and took it as a symbol to represent the town and the people who lived there. When there was a potential destructive threat to the Kneeling Nun rock formation in the course of mining operations, the locals grew concerned and acted to protect it. Many communities have strong ties to their own physical geography, symbolic features, and stories that connect the community to place, their own version of the Kneeling Nun.

PLACE AND SPATIAL POLICY

After having read the various spatial case studies presented in this book, we can draw some important conclusions for creating spatial policy across a variety of settings. Once you've done a thorough examination of the data and drawn conclusions, you can begin to thoughtfully consider lessons learned before acting to establish policies. Policies can be rules and actions, both formal and informal. A simple definition for policy is that policy is a framework for action. But the main thing to keep in mind when developing policy is that it should be data driven and accomplish something of specific value. A policy that is generated based on whim will not serve the greater population. A policy built on patterns identified in the data has a much better chance of meeting the needs of the local populations. Data-driven decision making is a common means to describe this process. In context of people, place, and resilience, this is an approach that, by necessity, requires careful consideration of the spatial component and how relationships and interactions affect what is happening within the boundary of the local area.

Quality data

It is important to make policy decisions that are rooted in valid, trustworthy, and well-sourced data. This means that the group of local decision makers should identify and have access to a range of relevant data in whatever forms are most appropriate to the question at hand. These may come in the form of primary data (collected yourself) and secondary sources (collected by someone else, perhaps for another purpose) including written, spreadsheet, database, and, of course, spatial data types. Data that captures local knowledge, skills networks, and abilities can significantly enable the community to achieve resilience. Data of local knowledge does not always exist, but rather needs to be identified and captured. Of course, in addition to the data itself, it is essential to consider the providence and quality of this data. For example, is the source of the data authoritative, from an official, trusted source? If data is not from a reputable source of known validity, one must be careful in how it is used. All data has potential value, but some may be of more value than others. As a rule, looking for the best available data, or considering alternatives, new collections, or other sources is always important. In the end, poor data sources will result in poor policy decisions.

When developing policy, it is equally important to be aware of local cultures and customs. Even the best-planned policies for a resilient community may fail if they are not feasibly rooted in local norms and practices. The norms and behaviors of communities in different places may vary greatly. In any given locality or community, there are typically diverse people, each with their own interests and practices. It is central to identify the different groups of people who are associated with the place of interest. Through direct interaction with the community, one can gain valuable knowledge history. But there is also tremendous value in learning about a place by spending time there.

In bigger cities, where you find multiple groupings within a named geographic region, it is important to pay attention to the nuances of smaller clusters or sub-groupings of people in a place. Units of analysis (city, village neighborhood, township) can all vary but can be useful in delineating where people live, who lives there, and culturally how they orient. In the larger community or region, any analysis must consider these multiple facets of the community and place. Often the one-size-fits-all approach fails to account for the cultural diversity of a community. Once groups are identified, one can then branch out to map them geographically and see how connections exist through social, economic, and political networks, which allow the community to function and sustain itself.

After identifying the culture of a place, the next step is to understand what is truly important to the local people. A good researcher will identify and understand variations in the local norms and values for a local populace. Different geographic regions will likely have different cultures that vary at multiple scales. It may be obvious that national cultures would be quite clear between France, China, Argentina, and Canada, each with different languages, foods, lifestyles, and so on. However, within a nation or region, we would also expect significant differences. Certainly, the local norms and culture of New York City, Dallas, Texas, and Los Angeles, California, are quite different, as are those between a large urban area and a small town. And, of course, even within a city, various neighborhoods can be quite distinct. As a good researcher, you want to capture and understand these variations within the local populace. This can occur through reading about the place, doing interviews, and doing surveys.

Symbolic nature of space and place

When creating policy, it's important to think about the crux of the local geography and the symbolic meaning behind those geographic features. What is the symbolic nature of space and place in that community? Earlier, we mentioned the symbol of the Kneeling Nun representing the rural mining communities in New Mexico. Similarly, in Northern California, there were giant redwood trees coupled with the symbol of Paul Bunyan, the mythical woodsman that represented the culture of people and place of the north coast of California. Another symbol of the local geography was Bigfoot, because a well-known video of Bigfoot was recorded near the town of Willow Creek, California, where Bigfoot sightings are a part of local lore. The town embraces this iconic symbol through a Bigfoot museum and an annual celebration of Bigfoot Days. Whether Bigfoot is real is not as important as the fact that the community chooses this as an icon to represent it. Such icons are typically rooted in something about the place, for example, something rural, something natural, or something wild.

This is a good place to think about the symbolic representation of a community. Sometimes a community will use multiple symbols, each reflecting the different populations that live there. Those symbols may appear as the local high school mascot, in the names of local businesses, or as images in the community serving to represent the community or place.

Build on local strengths

It is important to conduct a strengths inventory of the knowledge, skills, and abilities possessed by the local community. Who lives there? What are they good at? What are they passionate about? The people who live in a place and who have passion for that place will have the desire to do something to promote their local community, even in the face of hardships challenging their resilience. Every place, even those facing significant challenges to their resilience, will have things about it that are unique and special. For instance, in Silver City, New Mexico, a small mining town in the southwestern region of the state, local people did not have much money, but they had strong social networks. Residents came out to support their community at high school football games or at church, neighborhood barbecues, and festivals. The town was not rich, but it was committed to the place and would work together to help achieve success for people in the community. Being aware of a local community's strengths allows a policy maker to put effective community- and place-based ideas into play and help the community move forward.

FINAL THOUGHTS

As coauthors, we have been engaged in using GIS for the good of communities for more than a quarter-century. We see the value and the power of GIS to allow people in various places around the world to achieve the best of what they can be. In this time, these tools have become much more prevalent and accessible, not only to GIS professionals but to essentially everyone with an interest in exploring the spatial perspective. Our goal in writing this book was to share these stories by our contributing authors and to help readers develop an expanded sense of the power of GIS to address the difficult problems we collectively face in an ever-changing world. We hope that you have enjoyed exploring these case studies and that you will take away valuable ideas to help the communities and places you care about achieve spatial resilience.

Throughout this book, we have shared a diverse set of positive spatial stories that recount stories of resilience, be that preservation of traditional cultures and knowledge or protection of important resources essential to the social, economic, and environmental success of local communities, each with support from spatial thinking and GIS technology. Our overarching goal is to motivate other people, places, and communities with the recognition that they can achieve resilience using similar approaches in whatever context is most closely related to their own, while also learning from contexts and methods that may, on first glance, seem far removed. In an ever-changing world, every community will face challenges that test its resilience, be those changing environmental, social, political, or economic conditions.

While no one can precisely predict the future, a tremendous value of using GIS is the ability to use data to examine alternatives, tell stories, and visualize data in meaningful ways that can help to bring the spatial perspective to these complex questions in ways that can still be simple enough for all interested stakeholders to understand. Individuals who have passion, knowledge, and power and want to help their localities and the people who live there will ultimately rise to find a path to resilience. GIS has an important role to play in that larger equation by helping people see what resources they have and where they are located. Finally, GIS provides a big vision for action by layering valuable pieces of information together to see where gaps are located, where action is needed, or how policies can be instituted to manage and improve community resilience.

→

Contributors

(in alphabetical order)

MARIANA B. ALFONSO FRAGOMENI

University of Connecticut, Institute for Collaboration on Health, Intervention, and Policy, 2006 Hillside Road, Unit 4067, Storrs, CT 06269, USA; mariana.fragomeni@uconn.edu; +1 860-486-1947

Mariana Alfonso Fragomeni is currently working on a PhD in geography and integrative conservation, developing a collaborative heat response plan for the city of Savannah, Georgia, to understand possible barriers to incorporating climate knowledge in decision-making and land use planning. Her master's research looked at how urban design can impact climate, demonstrating how green infrastructure and building layout can impact ventilation and thermal sensation. As a graduate research assistant, Alfonso Fragomeni worked with the Georgia Sea Grant and the Carl Vinson Institute of Government to develop the St. Marys Flood Resiliency Project. Her research interests include resiliency planning, applied urban climatology, bioclimatology, collaborative GIS, and geodesign. Geographically, her professional experience and research have concentrated in areas of northeastern Brazil and Georgia, USA. Alfonso Fragomeni has a bachelor of science in architecture and urbanism from the Universidade Federal da Bahia in Salvador, Brazil, and an MEPD in environmental planning and design from the University of Georgia.

MAYLEI BLACKWELL

University of California, Los Angeles, César E. Chávez Department of Chicana/o Studies, Los Angeles, CA 90095-1559, USA; maylei@chavez.ucla.edu

Dr. Maylei Blackwell is an associate professor at UCLA. She has worked with the Indigenous social movements in Mexico for over 20 years. For the last 10 years, she has conducted community-based and collaborative research with migrant Indigenous communities from Mexico and Guatemala who live and work in Los Angeles. Her research expertise focuses on the intersection of women's rights and Indigenous rights in Mexico and California as well as the global struggle for Indigenous rights. More recently, she has documented cultural continuity and political mobilization among Zapotecs and Mixtecs from the Northern Sierra and the central valleys of Oaxaca, as well as the increasingly Mayan diaspora from Guatemala in Los Angeles. In addition, she is a noted oral historian and author of *¡Chicana Power! Contested Histories of Feminism in the Chicano Movement,* which was a finalist for the Berkshire Conference of Women Historians Book Prize and named by the Western History Association as one of the best books in Western women and gender history. Dr. Blackwell has served as an advisor to the Binational Front of Indigenous Organizations.

PAUL T. CEDFELDT

US Army Corps of Engineers, Portland District: GIS, CAD,
Mapping and Central Files Section, Portland, OR 97206, USA;
Paul.T.Cedfeldt@usace.army.mil; +1 503-808-4856

Paul Cedfeldt, supervisory physical scientist, is a certified geographic information systems professional (GISP) and chief of the GIS, CAD, and Mapping section at the US Army Corps of Engineers, Portland District. He has been working with geographic information systems since 1992. Since 1997, he has worked on diverse military and civil works projects across several Corps offices starting at the Cold Regions Research and Engineering Lab, then the Northwestern Division, and now at the Portland District. The primary focus of Cedfeldt's career with the Corps has been the application of GIS to support mapping, geospatial analysis, and data management. In his career, Cedfeldt has researched using GIS to identify functionally significant wetlands, applied GIS to help manage environmental restoration for the Army in Alaska, and led the GIS technical team for the Columbia River Treaty 2014/2024 Review. Cedfeldt earned a bachelor of arts in political science and environmental studies from Colgate University in 1992, and a master of science in natural resource planning from the University of Vermont in 1998.

SUE DAVENPORT

Kanyirninpa Jukurrpa, PO Box 504, Newman, Western Australia, Australia; sue.davenport@kj.org.au; +61 417-690-083

Sue Davenport is an advisory director to the board of Kanyirninpa Jukurrpa and a consultant to Puntura-ya Ninti (the Martu cultural knowledge program), which forms one division of Kanyirninpa Jukurrpa. Puntura-ya Ninti runs a suite of cultural programs, including the collection of oral histories and genealogies; the preservation of cultural and historical images, recordings, and artifacts; the cultural mapping of the Martu traditional lands; the writing of a history of Martu people since first contact; and the preservation of Martu language. Davenport has been working with Martu since 1987 and has a rich knowledge of Martu country, families, and cultural connections to country. She is a coauthor of *Cleared Out* (Aboriginal Studies Press, Canberra, 2005), a book detailing the first contact between one Martu group and Europeans, which took place in the Western Desert in 1964. She was a consultant to the 2009 documentary film based on that book, *Contact*. Before working with Martu, Davenport had a career in community development, including a term as national vice president of Amnesty International, during which she undertook a restructuring of the organization. She has a degree in anthropology and Aboriginal studies from the Australian National University.

JASON A. DOUGLAS

Department of Health Sciences, Crean College of Health and Behavioral Sciences, Chapman University, Orange, CA 92866, USA; https://www.chapman.edu/crean

Dr. Jason A. Douglas is an assistant professor of public health at Chapman University. He conducts research concerning social and environmental determinants of public health disparities in low-income communities of color. Working in a community-based participatory research context, Dr. Douglas collaborates with community-based organizations and underserved communities to develop an understanding of social and environmental factors manifesting in poor health outcomes at various geographic scales. These collaborative efforts and resulting findings are used to identify and implement empowerment pathways toward redressing social and environmental inequities and accompanying public health disparities.

ESTE GERAGHTY

Esri, 380 New York St., Redlands, CA 92373, USA; EGeraghty@esri.com;
+1 909-793-2853

Este Geraghty, MD, MS, MPH, CPH, GISP is the chief medical officer at Esri, developer of the world's most powerful mapping and analytics platform. She heads Esri's worldwide health and human services practice and is passionate about transforming health organizations through a geographic approach. Previously, she was the deputy director of the Center for Health Statistics and Informatics at the California Department of Public Health. There, she engaged in statewide initiatives in meaningful use, health information exchange, open data, and interoperability. While serving as an associate professor of clinical internal medicine at the University of California (UC), Davis, she conducted research on geographic approaches to influencing health policy and advancing community development programs. Geraghty is the author of numerous health and GIS peer-reviewed papers and book chapters. She has lectured extensively around the world on a broad range of topics that include social determinants of health, open data, climate change, homelessness, access to care, opioid addiction, privacy issues, and public health preparedness. She received her medical degree, master's degree in health informatics, and master's degree in public health from UC Davis. She is board certified in public health (CPH) and is also a geographic information systems professional (GISP).

MISHUANA GOEMAN

Gender Studies Department and American Indian Studies,
University of California, Los Angeles, Los Angeles, CA 90095, USA;
goeman@gender.ucla.edu

Dr. Mishuana Goeman, Tonawanda Band of Seneca, is an associate professor of gender studies, chair of the American Indian Studies interdepartmental program and associate director of the American Indian Studies Center at the University of California, Los Angeles. She received her doctorate from Stanford University's Modern Thought and Literature program and was a UC President's Postdoctoral Fellow at Berkeley. Her research involves thinking through colonialism, geography, and literature in ways that generate anticolonial tools in the struggle for social justice. Her book, *Mark My Words: Native Women Mapping Our Nations* (University of Minnesota

Press, 2013), was honored by the American Association for Geographic Perspectives on Women and a finalist for best first book from the Native American and Indigenous Studies Association (NAISA). *The Spectacle of Originary Moments: Terrence Malick's The New World* is in progress with the Indigenous Film Series, University of Nebraska Press. She has published in peer-reviewed journals such as American Quarterly, Critical Ethnic Studies, Settler Colonial Studies, Wicazo Sa, International Journal of Critical Indigenous Studies, Frontiers: A Journal of Women's Studies, Transmotion, and American Indian Culture and Research Journal. She has guest edited journal volumes on Native feminisms and another on Indigenous performances. She has also coauthored a book chapter in *Handbook for Gender Equity* on "Gender Equity for American Indians" and authored chapters in *Sources and Methods in Indigenous Studies* (Routledge, 2016), Macmillan Interdisciplinary Handbooks: *Gender: Sources, Perspectives, and Methodologies* (2016). Other book chapters include a piece on visual geographies and settler colonialism in *Theorizing Native Studies*, edited by Audra Simpson and Andrea Smith, (Duke University Press, 2014) and a chapter on trauma, geography, and decolonization in *Critically Sovereign: Indigenous Gender, Sexuality, and Feminist Studies* (edited by Joanne Barker, Duke University Press, 2017). She is also a codirector on a community-based digital community project, Mapping Indigenous Los Angeles (MILA), which is creating self-represented storytelling, archival, and community-orientated maps that unveil multilayered Indigenous Los Angeles landscapes. The created maps begin with the Gabrielino-Tongva and Fernandeño Tataviam while including those from diasporic Indigenous communities who make LA their home. The current phase develops curriculum for K – 12.

LAUREL HUNT

Los Angeles Regional Collaborative for Climate Action and Sustainability, University of California, Los Angeles, Institute of the Environment and Sustainability, Los Angeles, CA 90095, USA; lhunt@ioes.ucla.edu; +1 310-903-2316

Laurel Hunt is the executive director of the Los Angeles Regional Collaborative for Climate Action and Sustainability (LARC), which is housed at the UCLA Institute of the Environment and Sustainability. LARC is a network of key decision-makers ensuring a sustainable region prepared for the impacts of climate change, and it is one of six regional collaboratives in California supporting climate change science, policy, and planning efforts across sectors. Hunt is a technical expert and consultant for plans, policies, and events related to climate change, sustainability, land use, and transportation. Before joining LARC, Hunt served as the director of the Mediterranean City Climate Change Consortium (MC-4), a city-based network of over 100 international partners. She also has experience with the nonprofit sector in Los Angeles and both the US House of Representatives and the Center for American Progress in Washington, DC. Hunt received her graduate degree in urban and regional planning from UCLA.

PETER JOHNSON

Kanyirninpa Jukurrpa, PO Box 504, Newman, Western Australia, Australia; peter.johnson@kj.org.au; +61 419-270-931

Peter Johnson is the manager of strategy and governance at Kanyirninpa Jukurrpa and is an advisory director to the board of Kanyirninpa Jukurrpa. He oversees the development and implementation of strategic initiatives within the company, the governance of the company, and the Martu Leadership Program (MLP). The MLP is a community education and development initiative, covering areas such as corporate governance, trust administration, the criminal justice and penal systems, Native title processes, economic development, and adult education. Johnson began working with Martu in 2003 and was CEO of the company from its inception in 2009 until 2013. Johnson trained as a lawyer and worked in the community legal sector, specializing in Social Security law. He moved into business, starting a legal IT company that grew to 150 employees with offices in Washington, DC, London, Sydney, and Canberra. He has published in several areas, including legal expert systems, legal drafting methods, Aboriginal history and Aboriginal policy. He was coauthor (with Sue Davenport and Yuwali) of *Cleared Out* (Aboriginal Studies Press, Canberra, 2005). He has a bachelor of arts from the Australian National University and a bachelor of laws from the University of New South Wales.

REGAN M. MAAS

Center for Geospatial Science and Technology, College of Social and Behavioral Sciences, California State University, Northridge, Northridge, CA 91330, USA; regan.maas@csun.edu; https://www.csun.edu; +1 818-677-3515

Dr. Regan Maas is an associate professor of geography and environmental studies, *as well as associate director of the Center for Geospatial Science and Technology,* at California State University, Northridge. She has over 20 years of experience working in the geospatial industry and teaching in the geospatial sciences. She is well-versed in the myriad spatial analysis and visualization techniques relevant to geography, especially in areas of neighborhood and residential dynamics, health research, and spatial demography. Her research focuses on integrating GIS and geospatial science into public health and minority health disparities research and explores the various relationships between demographic characteristics, health outcomes, and the environment at the neighborhood level across the urban landscape. Her most recent work attempts to understand the varied ways resiliency is measured in minority neighborhoods across Los Angeles. Additionally, Dr. Maas has also been heavily involved in research looking at residential mobility and selection across socioeconomic groups, both in the US and abroad. Dr. Maas has a bachelor of science in biology from the University of Iowa, a bachelor of science in psychology from Iowa State University, a master of arts in geography from California State University, Northridge, and a PhD in geography from the University of California, Los Angeles (UCLA).

HEIDI P. MORITZ

US Army Corps of Engineers, Portland District: Hydraulic and Coastal Design Section; Portland, OR 97206, USA; Heidi.P.Moritz@usace.army.mil; +1 503-808-4893

Heidi Moritz, a hydraulic engineer, earned a bachelor of science in civil engineering from Valparaiso University in 1981. She received a master of engineering in ocean engineering from Texas A&M in 1991 in association with the US Army Corps of Engineers Coastal Engineering Education Program. Before working with the US Army Corps of Engineers (USACE), she spent two years working in Thailand on water resources projects for the Peace Corps. Moritz worked at the USACE Chicago District for eight years as a hydraulic engineer and at the Portland District for 22 years as a coastal engineer. Moritz's primary areas of work with USACE include navigation structure design, risk and reliability analysis of coastal structures, shoreline protection design, coastal structure design, and adaptation to climate change. Since 2008, she has been working in the Responses to Climate Change program, specifically to develop adaptation guidance to sea level change for Corps projects. Moritz is currently leading a national and international team in the development of an engineering technical letter addressing the analysis and impacts of the various components of total water level at a project site. She is also a member of the advisory board for the USACE Coastal Working Group.

HANS R. MORITZ

US Army Corps of Engineers, Portland District: Hydraulic and Coastal Design Section, Portland, OR 97206, USA; Hans.R.Moritz@usace.army.mil; +1 503-808-4864

Hans "Rod" Moritz has worked at the US Army Corps of Engineers (USACE), Portland District, for 15 years as a hydraulic engineer and the Chicago District for six years as a civil engineer. Between stints at Corps districts, he was employed as a consultant for two years. His working experience includes design and operation of coastal, maritime, and riverine infrastructure; management of waterborne sediment; application of risk and reliability techniques to optimize navigation and flood-risk reduction projects; evaluation of climate change vulnerability for water resource projects; data collection and analytic techniques dealing with oceanographic-estuarine-riverine processes; and computational fluid dynamics. Recent working experience includes contributions to the Columbia River Treaty 2014/2024 Review, Major Rehabilitation Evaluation for Mouth of the Columbia River, and USACE Asset Management framework for coastal navigation infrastructure. Moritz is a registered civil engineer in Oregon and is a member of the Committee for Tidal Hydraulics. He is a USACE subject matter expert for regional sediment management and climate preparedness-resiliency. Moritz earned a bachelor of science in ocean engineering from Texas A&M University in 1986 and a master of engineering in 1991.

BRENDA NICOLAS

César E. Chávez Department of Chicana and Chicano Studies, UCLA, Los Angeles, CA 90095, USA; bnico001@ucla.edu; http://www.chavez.ucla.edu/content/2014-cohort; +1 323-559-5897

Brenda Nicolas, Zapotec, earned her bachelor of arts at the University of California, Riverside, in sociology and Latin American studies. She holds two master of arts degrees: from UC San Diego in Latin American studies and from UCLA's Chicana/o Studies program. Her master's theses, "'Reclamando lo que es nuestro': Identity Formation among Zapoteco Youth in Oaxaca and Los Angeles" and "'Soy de Zoochina': Zapotecs Across Generations in Diaspora Re-creating Identity and Sense of Belonging," in urban and rural sites, examine how politically and culturally Zapotec youth and women maintain and (re)create their Indigenous identity and belonging against state notions of what it means to be Indigenous. Her dissertation looks at the three (first, 1.5, and second) Zapotec generations in California. Her research topics include settler colonialism, belonging, land, critical hemispheric Indigenous studies, and communal participation as forms of resistance and Indigenous futurity. Nicolas is a PhD candidate and staff member for the Latin American Indigenous Diaspora team of the Mapping Indigenous Los Angeles (MILA) project at UCLA.

KEVIN O'CONNOR

Department of Education, Mount Royal University, Calgary, Alberta, Canada; koconnor@mtroyal.ca; +1 403-440-8504

Dr. Kevin O'Connor is an associate professor in the Department of Education at Mount Royal University in Calgary, Alberta, Canada. He has taught in elementary and secondary schools and was an educational administrator for 15 years. Much of his current research and publications are based on the synthesis of multisensory pedagogy and interdisciplinary curriculum through the integration of experiential and place-based learning, science field studies, and Indigenous education. Dr. O'Connor is the national chair of the Canadian Association for Teacher Education (CATE)-Self-Study of Teacher Education Practices (SSTEP) group, serves as an editor and reviewer for numerous education research journals worldwide, and is as an advisory board member on the Canadian Research Institute for Social Policy (CRISP),

NASA-GLOBE International Science Education Group, European Scientific Institute (ESI), and the Paulo and Nita Freire International Project for Critical Pedagogy.

ALLISON FISCHER-OLSON

Head of Research and Community Outreach, Rainbow Bridge-Monument Valley Expedition—ONWARD!, Santa Monica, CA 90401, USA; afischerolson@onwardproject.org; onwardproject.org

Allison Fischer-Olson has a bachelor of arts in anthropology from UCLA. She joined Mapping Indigenous Los Angeles (MILA) as a staff researcher and coordinator after completing her master of arts in the American Indian Studies Interdepartmental program at UCLA. Her graduate work focused on community-centered collaboration in museum and archival spaces, as well as more broadly in issues surrounding cultural heritage. Through this work, she also took an interest in collaborative community storytelling projects using digital platforms, with MILA being a great project in which to explore this topic. She has worked at several museums, including the Fowler Museum at UCLA, where she put her graduate work to use, building experience working with collections and Native community stakeholders, building meaningful collaborative relationships, and addressing issues surrounding culturally appropriate representations. She currently works as the head of Research and Community Outreach for Rainbow Bridge-Monument Valley Expedition—ONWARD!, a nonprofit organization dedicated to bringing a historic expedition and its surrounding landscape to life using innovative and experiential approaches.

DEAN OLSON

University of Oregon, Department of Geography, Eugene, OR 97403-1251, USA; deano@uoregon.edu

Dean Olson started with Mapping Indigenous Los Angeles (MILA) as a staff researcher and GIS technician during his undergraduate work at UCLA in the geography department. His interest in representations of space, and an education in GIS and remote sensing, led him to questions about how maps represent space and the understandings they're designed to produce. As a graduate student at the University of Oregon, he has focused his research on the relationship between geopolitical representations and regional interventions in the polar north. While working toward his PhD in geography, Olson advocates for a meaningful engagement between critical geography and GIScience, mixed methods research, and advances in the fields of digital humanities and science and technology studies.

ROSANNA RIVERO

College of Environment and Design, University of Georgia, Athens, GA 30602, USA; rrivero@uga.edu; +1 706-542-6217

Rosanna Rivero is an assistant professor at the College of Environment and Design at the University of Georgia, where she teaches courses in GIS, environmental planning, and research methods. She has conducted research in wetland and coastal areas of Venezuela, Florida, and Georgia (US), among other projects. She collaborated with the Coastal Regional Commission of Georgia in identifying resilience strategies for the coast of Georgia, and has coordinated projects on geodesign on the coast of Georgia through a collaboration with Carl Steinitz and Hrishikesh Ballal, founders of GeodesignHub. Rivero has a bachelor of arts in urban planning from the Universidad Simón Bolívar in Caracas, Venezuela, and a master of science and PhD from the University of Florida.

MICHELE ROMOLINI

Center for Urban Resilience, Loyola Marymount University, Los Angeles, CA 90045, USA; michele.romolini@lmu.edu; +1 310-338-7443

As the director of research for the Loyola Marymount University Center for Urban Resilience, Dr. Romolini leads a team of professional and student researchers to conduct applied social and ecological studies to produce sound science that can support decision-making in the region. Her specific area of expertise is in examining urban natural resources governance and management using statistical, spatial, and social network analyses. Through her work, Dr. Romolini engages with municipalities, state and federal agencies, university collaborators, and nongovernmental organizations. She has a bachelor of arts in biology, a master's in environmental studies from the University of Pennsylvania, and a PhD in natural resources from the University of Vermont.

ALISON SMITH

University of Georgia, College of Environment and Design, Athens, GA 30602, USA; alisonls@uga.edu

Alison L. Smith is an assistant professor in the College of Environment and Design at the University of Georgia. She is a licensed landscape architect, an AICP certified planner, and has extensive experience with geographic-based technologies, with a focus on inventory, analysis, suitability, and the integration of GIS in the public design and planning process. Smith collaborated with the Coastal Regional Commission of Georgia, and Carl Steinitz and Hrishikesh Ballal, founders of Geodesignhub, on multiple geodesign projects focused on coastal resiliency at the county and regional scales. She has also integrated coastal resiliency projects into her design studio through a collaboration with the Georgia Conservancy's Coastal Program.

SHEILA LAKSHMI STEINBERG

Social and Environmental Sciences, Brandman University, Irvine, CA 92618, USA; ssteinbe@brandman.edu; http://www.brandman.edu; +1 949-585-2990

Dr. Sheila Lakshmi Steinberg is a professor of social and environmental sciences at Brandman University in Irvine, California. She joined Brandman University in 2013, and she teaches courses related to social and environmental sciences. Dr. Steinberg conducts research that focuses on space and place, and how research can benefit communities and the people who live there. She is committed to conducting applied research and encourages her students to use research to make a difference and to improve society through creating evidence-based policy based on sound research. Dr. Steinberg is also a prolific author and researcher. Her research focuses on the relationship of people, space, and place using a diversity of research methods. In 2015, she coauthored a book titled *GIS Research Methods: Incorporating Spatial Perspectives* for Esri Press and has coauthored a book for Sage Publications titled *GIS for the Social Sciences: Investigating Space and Place* (2006). Steinberg coedited *Extreme Weather, Health and Communities: Interdisciplinary Engagement Strategies* for Springer Press (2016). In 2016, she was the keynote speaker at the European Sociological Association (ESA) Qualitative Methods Research Network Conference in Krakow, Poland. Before joining Brandman University, Dr. Steinberg led the California Center for Rural Policy as the director of community research and led the master's program in public sociology at Humboldt State University in Arcata, California. She enjoys studying the intersection of culture, people, and place.

STEVEN J. STEINBERG

Los Angeles County, Internal Services Department, ITS Shared Services Branch, Downey, CA 90242, USA; ssteinberg@isd.lacounty.gov; http://isd.lacounty.gov; +1-562-392-7126

Dr. Steven Steinberg is the Los Angeles County geographic information officer (GIO). He oversees geospatial strategy for the County of Los Angeles, serving over 100,000 employees and 10 million residents, in collaboration with a team of highly skilled GIS professionals. He was first introduced to geospatial analysis technologies in the early 1990s as part of his master of science at the University of Michigan, and subsequently completed his PhD at the University of Minnesota as part of the team working on MapServer, one of the earliest web-based geospatial applications. From 1998 to 2011, he was a professor of geospatial science at Humboldt State University in Arcata, California. He was honored as a Fulbright Distinguished Chair (Simon Fraser University, Canada, 2004) and a Fulbright Senior Scholar (University of Helsinki, Finland, 2008). Following his academic career, he was the principal scientist and department head for information management and analysis at the Southern California Coastal Water Research Project, an environmental research agency, from 2011 to 2018, where he oversaw geospatial and data collection, management, analysis, and visualization systems and IT operations. Dr. Steinberg has been a certified GIS professional (GISP) since 2008.

ERIC STRAUSS

Loyola Marymount University Center for Urban Resilience, Seaver College of Science & Engineering, Los Angeles, CA 90045-2659, USA; Eric.Strauss@lmu.edu; http://cures.lmu.edu; +1 310-338-7337

Dr. Eric Strauss is a President's Professor of biology at Loyola Marymount University, with a specialty in urban ecology and wildlife studies. He is the founding executive director of the LMU Center for Urban Resilience (CURes) and founding director of Ballona Discovery Park, with a mission to bring the community and university closer together to develop and implement research, educational, and restoration programs. Professor Strauss was previously at Boston College, where he served for 15 years as the director of environmental studies, a faculty member in the biology department and director of a long-term study of coyotes in the Boston metropolitan area. With support from the National Science Foundation, Department of Education, and the USDA Forest Service, plus philanthropic foundations such as Annenberg, Gottlieb, and Collins, Dr. Strauss engages collaborative research specialties in animal behavior, urban ecosystem dynamics, and science education. A hallmark of his work is the inclusion of public participation in science and policy making. He is senior editor of the peer-reviewed journal *Cities and the Environment* as well as the senior author of *Biology: The Web of Life*. Dr. Strauss earned his PhD in biology from Tufts University in 1990.

WENDY GIDDENS TEETER

Fowler Museum at University of California, Los Angeles, Los Angeles, CA 90095-1549, USA; wteeter@arts.ucla.edu; http://pimu.weebly.com

Dr. Wendy G. Teeter is the curator of archaeology for the Fowler Museum, UCLA NAGPRA coordinator, and teaches periodically in the UCLA American Indian Studies program and the California State University, Northridge, anthropology department. She is on the Register of Professional Archaeologists and codirects the Pimu Catalina Island Archaeology Project, which seeks to understand the Indigenous history of the island and Tongva homelands through multidisciplinary and collaborative methodologies. Her interests, lectures, and publications focus on the protection and understanding of material and nonmaterial culture in the past, Indigenous archaeology, and the relationships between humans and the environment in North and Central America. She is also Co-PI for Mapping Indigenous Los Angeles (MILA), a community-based website devoted to storytelling through cultural geography and mapmaking as well as providing educational resources and curriculum. Dr. Teeter earned her PhD in anthropology from the University of California, Los Angeles, in 2001. She helped to develop the Tribal Learning Community & Educational Exchange Program in the Native Nations Law and Policy Center, UCLA

School of Law. She serves on a number of boards and committees, including as chair of the Society for California Archaeology Curation Committee and as an editorial board member of *Heritage & Society Journal.*

JACOB A. WATTS

US Army Corps of Engineers, Portland District: GIS, CAD, Mapping and Central Files Section, Portland, OR 97206, USA; Jacob.A.Watts@usace.army.mil; +1 503-808-4976

Jacob Watts is a geographer and certified geographic information systems professional (GISP) with the US Army Corps of Engineers, Portland District. He works to apply geospatial technology to a wide array of civil works and emergency management missions. In addition to performing GIS data management, analysis, and cartographic production, Watts specializes in remote sensing data acquisition from a variety of platforms and sensors, including satellite, manned aircraft, and unmanned aircraft systems as well as hydro-survey vessels. He develops integrated topographic and bathymetric elevation datasets as well as orthorectified aerial photo mosaics to support multiple engineering and environmental compliance activities. He has over 12 years of experience as a GIS analyst across a variety of public and private enterprises and holds a bachelor of arts in geography from Arizona State University (2006), as well as a graduate certificate in geographic information systems from Portland State University (2009).

Index